SAN MIGUEL
A Mexican Collective Ejido

SAN MIGUEL
A Mexican Collective Ejido

❧❦

RAYMOND WILKIE

STANFORD UNIVERSITY PRESS
STANFORD, CALIFORNIA
1971

*To the people of Mexico,
who have tried, by means of the collective
ejidos, to further the brotherhood
and true community of men;
and to my children, Larry, Kim, Pam, and Craig,
who I hope and believe will contribute
to the same goal*

Stanford University Press
Stanford, California
© 1971 by the Board of Trustees of the
Leland Stanford Junior University
Printed in the United States of America
ISBN 0-8047-0739-1
LC 79-119504

Acknowledgments

The 1953 fieldwork for this study was made possible by a grant from the Wenner-Gren Foundation for Anthropological Research. Additional fieldwork in 1966 and 1967 to update the material was financed by the Wenner-Gren Foundation and by the University of Kentucky Research Foundation. I am especially indebted to Ralph Linton and Wendell Bennett, who recommended me for these grants, and, at the Wenner-Gren Foundation, to Paul Fejos, the director in 1952, and Lita Osmundsen, the director of research in 1966.

It was my good fortune to have Sidney Mintz direct my research in 1952–53, when I was a graduate student at Yale University. I cannot fully express to him my appreciation for his intellectual stimulation, wise counsel, and friendship. To Clarence Senior, whose work first attracted my interest to the ejidos of the Laguna region, to Nathan Whetten, who also provided information on the ejidos, and to Charles Erasmus, who recommended this study to Stanford University Press, I would like to express my heartfelt gratitude.

Many Mexican government officials were helpful, and notably Ingeniero Victor Manuel de León, of the Ejido Bank in Torreón, who provided me with statistics on the collective ejidos that ultimately led me to choose San Miguel for study. When I returned to Mexico in 1966, I was fortunate indeed to have the assistance of Dr. Rodolfo Stavenhagen and Ingeniero Reyes Osorio, of the

Centro de Investigaciónes Agrarias. The staff of the American consulate (which was in Torreón until 1953) gave me every assistance, especially J. W. (Bill) Myers. Messrs. Rude and Curry of the American Department of Agriculture in Torreón were helpful in many ways in 1953, as was Robert Hardman, who was in charge of the department when I returned in 1966 and 1967.

I deeply appreciate the efforts of Ramón Sotomayór, of Torreón, who took the pictures of San Miguel; and those of Felipe Blanco Muñoz, who helped me with the maps. No editors could be more able, conscientious, and encouraging to an author than J. G. Bell and Barbara Mnookin, and I am grateful beyond measure for their assistance.

I will always be indebted to Nicolás Robles and his family for sharing their home in San Miguel with me with great hospitality and kindness. During my three visits the people of the community were unfailingly hospitable and friendly; to name those who helped me would be to name everyone I know in San Miguel, and many whose names I never learned. I must mention, however, Vicente Robles, Felipe Blanco, Fabian Robles, Florentino Reyes, and Aurelio Gonzalez, friends with whom I spent many enjoyable hours and who taught me much—not only about San Miguel and the collective ejidos, but about integrity, honor, and courage.

To name all those who have contributed significantly to this book is impossible; teachers, colleagues, students—all have played a role. But most of all I want to express my appreciation to my family, whose love, concern, encouragement, and forbearance have made this study possible and the work worthwhile.

R.W.

Contents

Introduction — xi

PART I. SAN MIGUEL TO 1953

1. Historical and Geographical Background — 3
2. Description, Demography, and Formal Organization — 25
3. Economic Organization — 54
4. Social Life and the Family — 90

PART II. SAN MIGUEL IN PERSPECTIVE

5. San Miguel Since 1953 — 131
6. San Miguel and Other Collective Ejidos — 149
7. Alternative Courses for the Collective Ejidos — 158

References — 183
Index — 187

Eight pages of pictures follow page 62

Tables

1 Distribution of Irrigated Land in the Laguna Region, 1928	12
2 Distribution of Laguna Land	22
3 Irrigable Land of Private Farms by Size of Unit	22
4 Characteristics of 361 Laguna Ejidos, 1948	23
5 Population of San Miguel, 1900–1960	32
6 Money Lent by the Ejido Bank to San Miguel for Wages and Other Expenses, 1953	63
7 Employment of San Miguel Libres, June 1953	67
8 Cotton Production and Yields in San Miguel and the Laguna Region, 1936–1952	76
9 Wheat Production and Yields in San Miguel and the Laguna Region, 1937–1951	78
10 Total Income of San Miguel, Including Money Spent Collectively, 1937–1952	79
11 Total Income of San Miguel Distributed to Individuals in Cash, Excluding Money Spent Collectively, 1936–1959	80
12 National Cost of Living Compared with Real Income of San Miguel, 1936–1967	81
13 Primary Sources of Income Variation Among San Miguel Ejidatarios, 1937, 1944, and 1950	82
14 Median Income of the Top, Middle, and Bottom Deciles of San Miguel Ejidatarios in Selected Years	83
15 Total Monthly Wages for Individual and Collective Work in San Miguel, 1953	85
16 Collective Work of San Miguel Libres, February 8–May 30, 1953	87

17 Total Income from Collective Work of San Miguel Libres, February 8–May 30, 1953	87
18 Weekly Employment of San Miguel Libres in Collective Work, February 8–May 30, 1953	87
19 Collective Work and Income of San Miguel Libres, June 21–27, 1953	88
20 Age Distribution of San Miguel Ejidatarios and Libres, 1953	94
21 School Enrollment of San Miguel Children, 1953	102
22 Rural and Urban Population of the Municipios of Matamoros and Torreón, 1921–1960	132
23 Wholesale and Consumer Price Indexes, Mexico City, 1936–1967	135
24 Total Operating Expenses of San Miguel, 1936–37 to 1959–60	136
25 Uses of Ejido Bank Loans to San Miguel, January–December 1960	137
26 Income Distribution of 5,000 Mexican Families, 1950 and 1957, by Percentiles	138
27 Income Distribution of 5,002,100 Mexican Families, 1956	139
28 Per-Family Average Monthly Income of San Miguel in 1950 and 1960 Compared with Income of Other Mexican Families	139
29 Total Monthly Wages Advanced to San Miguel in 1960 for Work in Cotton and Wheat	141
30 Number of Collective and Individual Ejidos in Mexico in 1953, by Location of Bank Branch	150

Introduction

The Mexican Revolution of 1910–21 was so clearly a struggle for political democracy and land reform that every presidential administration since has made at least token efforts to carry forward its goals, as defined by the Constitution of 1917. Although many Mexicans feel that the impetus of the Revolution has slowed, or stopped altogether, Mexico is a leader among Latin American nations, both in its land reform programs and in its attainment of a stable and democratic government.

Political stability has been achieved through the development of a powerful political party, the Party of Revolutionary Institutions (PRI), which has managed to incorporate most of its potential rivals. Political democracy has not fared quite as well; but all major interest groups are represented in the PRI, with the exception of the army and certain large financial and industrial corporations, which effectively exert their influence in other ways.

Land distribution, one of the major goals of the Revolution, has been accomplished in two ways: by ejido grants to individuals and by ejido grants to communities. Until 1934 all the ejido grants to communities were worked individually; they were usually used for subsistence crops, primarily corn and beans, and were associated with a very low degree of capitalization and productivity.

The Cárdenas administration of 1934–40 created a new kind of ejido grant, with land to be held and worked cooperatively. By this means the government sought to institute modern agricultural methods and to avoid the disadvantages of distributing land in very small parcels to peasants lacking both capital and technical knowledge. Over 300 of these collective communities were established in the Laguna region of Coahuila and Durango alone, and about the same number were set up in five other areas of Mexico. By 1953, when I first undertook this study, this immense social experiment had been in operation for almost two decades.

The ejidos are the core of Mexico's land reform program and a major factor in its political and economic stability. Yet in spite of their importance to Mexico and their potential importance as a model for other countries, there are remarkably few studies of their success as economic organizations. Even fewer studies have been concerned with the effects of the ejido economy on the noneconomic aspects of community life.

Of the more than 20 anthropological and sociological studies of Mexican communities published between 1925 and 1953, the only one that presents a detailed description and analysis of the economic system is Oscar Lewis's *Life in a Mexican Village: Tepoztlán Restudied* (1951). Lewis's conclusions about the ejido economy in Tepoztlán sum up and are consistent with the findings of anthropologists, sociologists, and Mexican agriculture officials regarding the individual ejido grant. Of Tepoztlán's ejido economy he writes: "The ejido has not really solved the land problem ... due in part to the small amount of land that could be distributed" (p. 128). It is his conclusion that one of the crucial problems is the rapid population increase with no accompanying increase in resources or improvement in production techniques. In his words, there has been "a return to a more primitive type of production (hoe culture on the hills and scrub forest) in an effort to escape the devastating effects of a money economy during a period of inflation" (pp. 9, 157).

Apart from Lewis's study and despite the uniqueness of the ejido experiment as a source of data for social scientists interested in the process of planned change in underdeveloped countries, I found that virtually nothing had been written on developments after 1940, and nowhere was there an anthropological case study of a single ejido collective community, though Nathan Whetten's *Rural Mexico* (1948), and Clarence Senior's *Land Reform and Democracy in Mexico* (then an unpublished manuscript) contained excellent general descriptions of the Laguna region's collective ejido system.

Intensive case studies of specific collective ejido communities are vitally needed to supplement the picture provided by regional and national statistics, which do not necessarily give an accurate picture of the economic life of individuals or communities. Nor is the quality of life in its noneconomic aspects at all clear from such data. For example, in the decade immediately preceding the Mexican Revolution, the vast majority of rural Mexicans had neither economic nor political rights; in fact, some were still held in virtual slavery. Yet most economists of the period, foreign and Mexican alike, described Mexico as economically prosperous. This description was based on gross national indices of production, commerce, construction, and income, indices that ignored the life of most of the people.

To note the misuse of large-scale national statistics is not to deny their usefulness, but it does point to the need for interpreting such data in combination with qualitative studies of individuals and communities, prepared by a social scientist who lives with the people he is studying, who participates in their community activities, and who shares for a while their hopes and joys, their work and frustrations. With such personal and immediate experiences for reference, national and regional statistics take on a very different meaning. It was with this in mind that I undertook to describe the changing economy and social organization of a single collective ejido community.

Because the Laguna cotton region of northcentral Mexico was the first collective ejido area established, and also the largest and most economically important one, I chose it for my study. Of the 300-odd collective communities in the area, I selected the ejido of San Miguel because it was in the middle range in total population (1,200), in area of cropland (four hectares per *ejidatario*), in the number of charter ejido members (150), and in degree of mechanization. I also chose San Miguel because, in the drought year of 1953, when many ejidos were plunged into economic crisis, its six wells and its canals, which connected with both the Nazas River and the Aguanaval River, provided an assured water supply. I soon realized that San Miguel was not as "typical" as I had thought. In fact, it was one of the most successful and important ejidos in the Laguna region, famous not only for its economic advances but also for its role in the Central Union of Collective Ejidos. President Cárdenas had visited the ejido twice, the noted artist Diego Rivera once; its officials went often to Mexico City to plead the cause of the Laguna ejidos with high government officials. Still, what information may be lost because San Miguel proved to be not completely "typical" is more than compensated for by the fact that its life has been so closely related to key decisions and decision-makers at the regional and national level.

The people of San Miguel, like the majority of Mexico's rural population, are farmers and rural workers. They are socially, economically, and politically marginal to the larger urban society. But there is a significant difference between the rural proletariat of the Laguna region and the peasants who constitute the population of most of rural Mexico. The collective ejido communities are highly capitalized, cash-crop businesses. They represent Mexico's major effort to meet the immediate economic needs of the rural population and at the same time increase agricultural productivity by the application of knowledge and techniques that require capital and efficient organization.

Critics of the collective ejidos contend that the government has merely replaced the hacienda owners, that the ejidatario is no better off than the hacienda peon, and that agricultural production has suffered as a result of the government's bureaucratic inefficiency. Since 1940 these critics have been influential in the federal government, which has considered the collectives more a problem than a solution. The creation of new collectives has ceased, and those already established have received only grudging support from the government. Indeed, the government has provided more and more assistance to the private farms *(pequeños propiedades)*, which are consequently able to institute the kind of intensive farming methods that lead to increased productivity.

The struggle between the rural workers and the private estates has thus taken a new turn in this century. Although the rural workers now have half the land, they do not have the financial resources, knowledge, and organization to compete effectively in the world market. The individual and communal ejidos continue to operate at the subsistence level; the collective ejidos are dependent on the government for any successes they have in the world market. The competition between the private landowners and the twentieth-century descendants of the Indian peons is now not primarily for soil, but for the resources that must be added to the soil to make it productive. This study of a collective ejido is a description of that struggle.

For five months in 1953 I lived in the household of Nicolás Robles, one of the younger members of the ejido. Besides Nicolás himself, the household was made up of his wife, his four children, his non-ejidatario cousin, and the cousin's wife and child.

Nicolás was my own age, 27, and since I too had four young children, I readily identified with him. During my stay in San Miguel, I visited in every household, interviewing and observing, but it was the Robles family that I became a part of, and on whom I relied for subjective validation of generalizations about San Miguel and other collective ejidos. The processes and trends in the collective ejido communities of the Laguna region and in the larger Mexican society became specific and real to me through my personal experiences in the Robles household; each of the community characteristics and processes I describe in the following pages has its counterpart in the lives of Nicolás and his family.

My hypothesis when I went to the Laguna region in 1953 was that the community-owned and -operated ejidos, if economically successful, would result in a more cohesive and stronger community organization. I anticipated that the educational, social, and legal systems would function more effectively than they had when the community was part of a privately owned cotton plantation. I hypothesized also that individual families would be strengthened as a result of their increased and stabilized incomes. These hypotheses were confirmed by observations, interviews, and statistical data collected in 1953, 17 years after the ejido was established.

I also found three major developments that I had not anticipated: an increased variation in income and living standards among the ejidatarios; a marked population increase; and the individualization, or decollectivization, of important aspects of the economy, including property rights, work methods, and income. In addition, I found that the formal ejido organization provided little flexibility for the growing population of non-ejidatarios and scant opportunity for them to fit into the economic structure. Nor was there enough work outside the community to accommodate the growing population.

On returning to San Miguel in 1966 and 1967, I discovered that these trends had continued, and that, in addition, the ejido had split into two conflicting political and economic groups. Similar divisions and conflicts had already occurred in many less successful ejidos since 1940. Indeed, in some cases the Ejido Bank had encouraged these divisions, and had lent money only to those ejido sectors it considered good financial risks. In some ejidos conflict grew out of a struggle between the labor agency of the government, the Confederación Nacional de Campesinos (CNC), and the nongovernment, pro-Communist organization, the Central Union of Collective Ejidos. The conflict was also caused in part by economic problems arising in the doubling or tripling of the ejido populations and the inflexibility of government regulations that prevented the adjustments and transfer of ejido rights necessary to accommodate the growing non-ejidatario population. Recently, there have been some indications that the Mexican government may, in the future, give more guidance and support to the collective ejido communities than has been the case to this point.

The data for this study were obtained from my own observations in 1953, 1966, and 1967, from historical records, and from interviews with residents who lived in San Miguel (some in the early 1900's) when it was a hacienda. The historical and geographical background of the land reform of 1936, in which the collective ejidos were created, is described in the first chapter. Following that is a description of San Miguel from 1936 to 1953, covering economic, political, religious, educational, leisure, and kinship aspects of community; and a chapter on San Miguel from 1953 to the present. My major focus in this description is on the ways the ejido community has been modified by demographic and economic changes. In the last two chapters I summarize the studies of other Mexican collective ejidos and discuss the alternative paths that are open to the federal government—and to the ejidos themselves.

Part I
San Miguel to 1953

I
Historical and Geographical Background

Geography of the Laguna Region

The Laguna region of northcentral Mexico is located in the southwestern corner of the Bolsón of Mapimí, into which the Nazas and Aguanaval rivers drain.* It comprises the *municipios* (townships) of Torreón, Matamoros, San Pedro, Viesca, and Francisco I. Madero, in the state of Coahuila, and the municipios of Lerdo, Gómez Palacio, Mapimí, and Tlahualilo, in the state of Durango. Together these municipios incorporate about 1.6 million hectares, of which only about 10 per cent is cultivated in an average year because of the limited water supply. The region has a mean elevation of about 3,000 feet above sea level and is shaped like a leaf, with the Nazas River representing the stem and the canal system the branching veins.

The Laguna region is not a "natural" geographical division, and does not differ in climate or topography from the surrounding desert. But viewed from an airplane, its boundaries are straight, with clean lines separating the gray desert scrub vegetation from the green rectangular fields of irrigated cropland.

Climate, Soil, and Water. In this hot desert climate, the mean annual precipitation, based on statistics for the years 1921–47, is 10 inches, almost half of which falls in a four-month period. With a rate of evaporation ten times the rate of precipitation, agriculture is impossible without irrigation. The mean temperature of

* The data on location, climate, soil, and water are from the Secretaría de Recursos Hidráulicos (1951), Chaps. 2, 4, 5.

the coldest month, January, is 42.4° F., that of the hottest month, June, 92.8° F. The relative humidity ranges from 40 per cent to 60 per cent throughout the year, so that with a slight breeze the hottest days are not uncomfortable except in the direct sun.

The seasons are not as well marked as in lower altitudes, but there is enough of a difference to limit the growing season of cotton, the major crop, to nine months. The winter season of frosts and occasional light snow lasts from about mid-November until March, though frosts have occurred as late as April. Plants like cotton, which are not frost-resistant, cannot be grown during these months; accordingly, this season is used for irrigating and preparing the soil for planting. In addition, wheat, which is not susceptible to frost damage, is planted in the late fall, a staggering of crops and growing seasons that permits more efficient use of the water supply.

June, July, August, and September are the months of greatest precipitation. The rain is sometimes sufficient to help the maturing cotton and corn, and is occasionally early enough to injure the ripened wheat, but it is unpredictable both in quantity and incidence, and there is little the local farmers can do to take advantage of it.

Except for some areas of sand and alkaline soil, the land of the region is suitable for agriculture, though it is deficient in important elements, notably nitrogen, phosphorus, manganese, and certain organic materials. Agricultural experts have recommended the use of leguminous plants, to be turned under for fertilizer, and the addition of nitrogen to the soil; they have also recommended that saline soils not be cultivated. Most of the soil of San Miguel is of good quality but its productivity would almost certainly be increased if these recommendations were fully implemented.

The most crucial factor in the regional economy is the water supply. The Nazas and Aguanaval rivers and wells are the principal sources of water. The average amount of land irrigated annually by each between 1917 and 1947 was as follows: Nazas River, 96,000 hectares; wells, 50,000 hectares; Aguanaval River, 10,000 hectares. The annual variation in available water is very great, however. The median annual flow of the Nazas (in the same period) was 125 million cubic meters, with a range of from two million to 861 million cubic meters; the median annual deviation from the mean was 30 million cubic meters. The Aguanaval is much less important, and even more variable from year to year.

The Nazas flows intermittently in August, September, and October, the Aguanaval, when it flows at all, in June and July. Before the Nazas was brought under control with the completion of El Palmito Dam in 1944, the primary means of irrigation was deep flooding. Once most of the Nazas waters could be stored in the reservoir behind the dam, the method of irrigation changed over to a shallow initial flooding with two or three auxiliary irrigations during the growing season. The earlier method wasted water but brought more sediment to the fields and did not deplete the soil of necessary minerals as rapidly as the new system. As yet no storage reservoirs have been built to control the Aguanaval's flow. The drainage area of the Nazas covers some 13,000 square miles and extends 200 miles westward to the crest of the Sierra Madres; the drainage area of the Aguanaval covers 9,000 square miles and extends 200 miles southward to the city of Zacatecas.

Since the 1930's wells have furnished one-third of the region's irrigation water in normal years, and as much as two-thirds in periods of drought. The water table is dropping, however, and there is a danger that wells may be of little use in a decade or two. The situation was already critical in the 1950's; many wells were dry, and the government had forbidden the construction of new ones except in case of extreme emergency. Moreover, much of the well water is bad. For agricultural use it has been classified this way: bad, definitely should not be used, 26 per cent; doubtful, probably should not be used, 20 per cent; saline but can be used, 43 per cent; good, not saline, 11 per cent. Use of the two worst classes as the lone source of irrigation water would ruin the land in less than ten years. However, since irrigation is done with well water alone only in drought years, the situation is somewhat better than it might seem.

As of 1953, the total area that could be irrigated annually was about 150,000 hectares, an amount far below the area the ejidos and the *pequeños propietarios* (small landowners) are legally entitled to irrigate. There is thus a continual source of conflict inherent in the distribution quotas: if the Federal Water Commission interpreted the laws to give priority to the ejidos, the pequeños would receive no river water at all, except in years of unusually heavy flow. In practice, the commission allocates water to the ejidos and pequeños in proportion to their total irrigable area. This means that the ejidos have never been able to irrigate all of their available land.

Transportation, Communications, and Settlement Pattern. The Laguna region is one of the few parts of Mexico whose development was greatly influenced by railroads. In 1880 the region was traversed by a main north-south railroad, which was crossed at Torreón by an east-west line eight years later. The main north-south route through central Mexico had previously connected Mexico City to Santa Fé, by way of Querétaro, Zacatecas, Durango, and Chihuahua, a route that lay some miles to the west of the Laguna region. Once the new lines were laid, Torreón, which was smaller than Lerdo or Gómez Palacio, grew rapidly, becoming Mexico's fifth-largest city and three times the size of its sister cities together.

In addition to the main railroads, there are almost 100 miles of interurban track in the region. But since the development of a good system of paved roads, and with the ever increasing use of buses and trucks, railways have become relatively unimportant, both within the region and as a connection with other parts of Mexico. Railroads were not an important means of travel or transport for San Miguel.

The main road of the Laguna region follows the meandering course of the Nazas River from Torreón, but in 1936 a straight new highway was built connecting Torreón more directly with the state capital, Saltillo, via Matamoros. Since then, many smaller farm roads have been built, and today there are few communities in the region that are more than an hour away from an urban area. Trucks and buses are an important means of transportation. Most ejidos have at least one truck and almost all of them are close to an interurban bus line.

Three bus companies operate between Torreón and the outlying towns. In 1953 a bus passed San Miguel every 20 minutes from early morning until late in the evening. According to bus company records, almost one-fourth of the ejido's population went to the city daily. The ease of transportation by bus has had important effects on the community. For one thing, no artisan class has developed in San Miguel, since the stores of Torreón and Matamoros are only minutes away. For another, travel to other regions for work or, in exceptionally good years, for vacations and religious pilgrimages has been greatly facilitated. Thanks to the improved means of transportation, the trip between San Miguel and the United States today is probably easier than the 200-mile journey between San Miguel and Saltillo, the state capital, was in hacienda days.

The Laguna region is also linked to the outside world by airlines, which have daily flights to the north, south, east, and west. For the rural workers, air travel has no direct importance. However, many government and business officials with whom they are in contact use airplanes frequently, some even commuting weekly from Mexico City.

Travel within the ejido and to nearby communities is primarily by foot or by horseback or donkey. It is seven kilometers from one end of San Miguel to the other, which means a daily trip to work of half an hour or more for many of the ejidatarios. Visiting between adjacent ejidos is common, both for formal and for informal gatherings, and travel is not difficult. When a number of people are going to a fiesta or some other community activity, the ejido truck is used.

There is no post office or regular mail delivery in San Miguel; in any case, the amount of written personal communication is small. The community's mail is delivered to a post office box in Matamoros and collected at irregular intervals by one of the ejido officials. A week's mail for the entire community usually consisted of no more than 25 pieces, a good part of which was concerned with official business. Much of the personal mail was to relatives outside the Laguna region, often invitations to come to the ejido and work during the cotton-picking season.

The radio and Torreón's two major newspapers, *El Siglo* and *La Opinión,* keep the members of the community informed, though the interpretation of current events most readily accepted in the ejido is that given by the leaders of the Central Union of Collective Ejidos, which San Miguel strongly supported. Almost all ejidatario families have radios, though few listen to them regularly. There are few telephones, not because they are too expensive but because there is little need for them.

Thanks to the excellent transportation and communication system of the Laguna region, the people of the ejidos are not limited to community institutions and facilities. The solidarity and homogeneity of isolated communities that are enforced by the absence of alternative behavior patterns are not found in San Miguel. Most individual and group activities are chosen from among several alternatives. San Miguel, like the other ejidos, feels itself a part of the region, the nation, and the world. Many of the people are vitally interested in events in these larger societies. The world prices of cotton and wheat are naturally of great concern, but so are the political struggles in other countries, for the ejidatarios are fully aware that the fate of Mexico is linked to that of the rest of the world. The acculturation of the rural population to urban dress and urban ways has been rapid since 1936, facilitated both by the transportation and communication systems and by the growth of personal income. Many of the rural inhabitants have relatives in the cities with whom they maintain close contact. The members of the ejido communities do not feel that they are different from urban dwellers, though they recognize the social and economic gulf that lies between them and the upper classes of the region.

Land use, the canal system, and the railroads all helped determine the settlement pattern of the Laguna region. The internal structure of the cities was largely based on the Spanish tradition of building towns around a square central plaza, but the ideal pattern was modified in Torreón and other Laguna cities, first by the railroads, then by roads and highways. Torreón, for example, is laid out as a long, thin rectangle, only eight or nine blocks in width, stretching for several miles along the railroad track. In most of the smaller towns and villages, the development has been the same: businesses and residences string along the arteries of transportation, and the plazas are virtually deserted except on fiesta days.

Most of the hacienda communities had a *casa grande*, or main residence, at one end of a square, with the peons' dwellings lining the other three sides. In a few of the most prosperous haciendas, the central area developed into a plaza, but in most rural communities the central square remained bare. The size and shape of the haciendas, as well as the location of the canals, largely determined not only the size of the rural communities but also their relationship to one another. The number of people in a given area depended primarily on the amount of irrigated land, but whether they were divided into many small communities or into a few large ones depended on the size of the hacienda to which they were attached and the work force the landowner needed. Essentially, then, the Laguna region, unlike many areas, does not have a settlement pattern primarily defined by geography; rather, it is an area that has been shaped to a large extent by its socioeconomic organization and the ways in which technology has been used to exploit the natural environment.

History of the Region

The Colonial and Hacienda Eras. The Lagunero Indians, who inhabited the Laguna region before the Spanish Conquest, were apparently a Nahua-speaking people, more closely related culturally to the Aztecs than to the surrounding Plains Indians. They practiced some agriculture on the banks of the Nazas and Aguanaval rivers and around the two lakes into which the rivers drained (Beals 1932: 99). If the population estimates of Jesuit priests in

1598 are correct—they calculated 16,000–20,000 inhabitants, an estimate some feel is too high—the Laguneros must have obtained a sizeable proportion of their food from agriculture, for it is difficult to see how the desert could have supplied enough game and wild plants for so many people.

The first Spaniards to come to the region were the troops of Captain Francisco Urdiñola, who began establishing military garrisons in northern Mexico in 1582. These military conquerors were soon followed by the Spanish priests. The Lagunero Indians required no "pacifying" at first, but when their lands were encroached upon they revolted and were decimated by Spanish soldiers; smallpox also took a heavy toll. Apart from Urdiñola himself, who carved out an estate of more than a million acres, the Jesuit priests were virtually the only settlers in the area for the next 100 years. They built a mission at Santa María de las Parras in 1600, bringing in Tlaxcalan Indians to work the land and to serve as a buffer against the local Indians.

The area offered no inducement to small freeholders throughout the seventeenth century because of the possibility of Indian attack; by the eighteenth century occupation by small-scale settlers was forbidden by the owners of the huge estates that had developed. In 1731 the Laguna region was bought from the Spanish Crown by the Marqués de Aguayo, a descendant of Urdiñola. At this time the region was known as Shepherd's Corner and was used only for cattle ranching. A few squatters, possibly the remaining Lagunero Indians, continued to live along the river banks, though the Marqués explicitly forbade anyone but his employees to live on his property.*

The only small freeholders in the region in the early eighteenth century were under the protection and domination of the Jesuits. When the Jesuit lands were confiscated in 1787, several enormous estates were established, with the result that free settlement was even further discouraged. But the breakup of these large estates

---

* According to the historian Miguel de Mendizabel (1964: 225–70), the Marqués declared: "Better that all my cattle perish from the arrows of the Apaches than that I give one foot of desert to these intruders."

History and Geography

began in 1824, when the Republican Constitution abolished the *mayorazgo* (entailment) privileges of the titled noblemen. In that year creditors got a legal judgment against the then Marqués de Aguayo and took over the management of his estate. Between 1824 and 1848 the vast cattle ranch was broken up into several large estates, one of them incorporating the entire Laguna region.

The region began its modern capitalistic era in 1848, when Leonardo Zuloaga and Juan Ignacio Jiménez, joint owners of most of the land in the region, built dams across the Nazas River at the place where Torreón now stands. Disputes over the control of the water soon began, both among various hacienda owners and between them and the downstream settlers of Matamoros. These struggles increased in intensity, culminating in a battle between private armies in 1862. The townspeople won the battle, but the struggle continued, ending only after President Benito Juárez intervened in 1864 to give Matamoros title to its land and the right to half the water of the Aguanaval River.

Following the death of Zuloaga in 1865, the division of the Laguna region continued rapidly, in part because of the financial ineptness of his widow and in part because of the need for capital to intensify agricultural operations. Zuloaga's widow borrowed almost 30,000 pesos from two men, Guillermo Purcell and J. S. O'Sullivan, and used one of her haciendas, El Coyote, as collateral for another loan. Both transactions resulted in the breakup of Zuloaga's estate, a process that continued after his widow's death.

The initial phase of growing capitalist agriculture, in which land was subdivided to obtain additional capital, was followed by a period in which the dominant sources of capital were foreign stock companies, with agricultural operations in the hands of administrators rather than individuals or families. The largest of these was the Tlahualilo Land Company, organized in 1885 by Mexicans but soon taken over by a British firm. The Tlahualilo company bought 110,000 acres of Laguna land, together with the rights to enough Nazas River water to irrigate all its holdings. These water rights were immediately contested by downstream landowners, and a long and complicated court battle ensued.

Neither side won a clear victory, and in the end a government commission was set up, which has regulated the distribution of irrigation water since (until 1936, largely on the basis of the *status quo ante*).

Though the Mexican Revolution began, in a sense, in the Laguna region, the birthplace of Francisco I. Madero, and was fought primarily for agrarian reform, the land tenure system in the region remained fundamentally the same after the Revolution. True, some Spanish landowners were replaced by Mexicans, but all this really meant was a change of personnel. As Table 1 shows, 32 per cent of the irrigated land was in the hands of six persons in 1928, and 80 per cent was owned by 45 estates. The land concentration was probably even greater than indicated, for one man or one land company often owned several properties, each of which was counted as a separate unit in the censuses.

Characteristics of the Laguna Haciendas. In most of Mexico (and in government censuses), the term hacienda was applied to holdings of 10,000 hectares or more. However, because of the intensity of agricultural production in the Laguna region, estates of half this size yielded a comparable income and were commonly called haciendas. The Laguna hacienda, the most important social and economic unit in the region from 1848 until 1936, also differed materially in its economy and social organization from the typical hacienda as described in three of the standard works on the subject: *The Land Systems of Mexico,* by George McBride, *The Mexi-*

TABLE 1
Distribution of Irrigated Land in the Laguna Region, 1928

Size of property (hectares)	Number of properties	Total area (hectares)	Per cent of total area
1–100	42	2,100	1%
101–500	65	19,500	7
501–1,000	34	25,500	10
1,001–5,000	39	117,000	49
5,001–10,000	1	7,500	3
Over 10,000	5	70,503	29
Total	186	242,103	100%

SOURCE: Guerra Cepeda 1939: 13. Columns in this and subsequent tables do not add up to 100% because of rounding.

can *Agrarian Revolution,* by Frank Tannenbaum, and *Rural Mexico,* by Nathan Whetten. The differences can be summarized as follows:

1. Capitalization. The Laguna haciendas were highly capitalized. The sums needed to build and keep up the irrigation systems, to purchase the machinery, tools, and draft animals required for intensive agriculture, were beyond the reach of most of the landowners, who had to rent their land to companies to secure additional operating capital. The largest and most highly capitalized haciendas in the region were owned by foreign stock companies. The typical Mexican hacienda had little working capital. The owners tried to make their estates self-sufficient and to pay their workers in perquisites or keep them attached to the hacienda by debt peonage. Technology was relatively primitive and required little capital.

2. Administration. Most of the Laguna property (if rated by value) was administered by the renting land companies or by the employees of stock companies; only a few were operated directly by an individual owner or his surrogate. The typical Mexican hacienda was, in Tannenbaum's words, rooted in "absentee ownership, indirect management, and security of income [and] the property was managed through an administrator" (1929: 105).

3. Payment of Workers. The Laguna workers received a daily wage as well as certain perquisites, notably a dwelling and (in San Miguel) the privilege of grazing a few animals. They did not have subsistence plots and were not attached to the soil by debt, largely because the great fluctuations in the labor needs of the hacienda made it desirable that the workers be free to come and go; the haciendas wanted to be responsible for as few permanent workers as possible. Workers in the traditional Mexican hacienda were unremunerated, were paid in kind, or were obtained through indirect tillage arrangements (Tannenbaum 1929: 126–27).

4. Sources of Supplies. Nearly all of the supplies of the Laguna haciendas, including staple foods, were purchased outside the hacienda. Barnyard animals and fowl were raised within the hacienda to provide milk, meat, and eggs, and the adobe blocks needed for hacienda buildings were produced there as well. Apart from this,

the peons bought food and clothes with their cash wages, though the *tiendas* (stores) extended credit and collected directly from the hacienda. The typical Mexican hacienda aspired to economic self-sufficiency. Food, clothing, and tools were manufactured within the hacienda, which, ideally, had its own woodlands, grazing land, cropland, and sources of water. These haciendas had a full complement of occupational specialists to manufacture or provide the necessities of life.

5. Type of Crop. The principal crop of the Laguna haciendas was cotton, a cash crop whose value was determined in a world market; it was not used within the hacienda. The secondary crop, wheat, was too expensive for the peons to use as a staple food; it too was grown as a cash crop. The main crop of the typical Mexican hacienda (with some regional variations) was corn, the staple food of the peons. Since it was produced within the hacienda for hacienda use, world market conditions could not reduce the peon's income below the subsistence level, though the hacienda owner's profit was susceptible to great fluctuation.

6. Relationship Between Peons and Hacienda Owners. In the Laguna haciendas, this relationship was based on the exchange of labor for a daily wage. The conditions of employment were determined by the hacienda administrators, who not only had economic power but also maintained political control of the municipio administration. (In the case of San Miguel, the hacienda owner was the municipio president.) In the typical Mexican hacienda, the peon had no work alternative; he was completely dependent on the hacienda owner. Ideally and often in fact in this paternalistic system, the peons had an emotional attachment to the owner, regarding him as their benefactor and protector; he in turn gave them small gifts and lent them money as a personal favor. In describing this relationship, Whetten writes (1948: 101): "The occasional visits of the *hacendado* and his family to the hacienda, usually during the harvest or the planting season, were accompanied by a certain display of paternalism toward the resident population. A few small trinkets for the children of the peons were brought along, a few coins were distributed here and there,

a local fiesta was arranged, and the *hacendado* was regarded more as a distinguished visitor than as an active participant in the enterprise."

7. Class Structure. Only two classes were represented in the Laguna haciendas: an upper class, made up of the owners and administrators, and a lower class, made up of the peons and the slightly more privileged *mayordomos* (work chiefs). There were no marriages or social activities between these classes. No middle class existed within the community; the small shopkeepers, businessmen, and political officials were all in the urban centers. The Laguna hacienda community was thus essentially only a geographically segmented part of a larger social structure. The classic Mexican hacienda had a complete social structure in that it incorporated a middle level of occupational groups consisting of artisans, small tradesmen, civic officials, schoolteachers, and all other specialized workers needed to meet the requirements of the society. "The hacienda usually contained the essential supply of services characteristic of an independent community—a store, a church, a post office, a burying-ground, a jail, and occasionally a school. The houses and farm buildings . . . were all constructed of local materials by local personnel, and workshops were maintained for the making of tools, implements and other essentials" (Whetten 1948: 100).

8. Political Organization. The Laguna hacienda was not a political unit; only one civil position was represented there, a very minor official who was appointed by the municipio president. The hacienda owners together exercised control of the municipio government and were often themselves the political officials. The typical Mexican hacienda often constituted a political unit, with its own government officials and police. Where this was true, the hacienda owner was in direct political control and did not have to cooperate with other landowners in the joint administration of a municipio.

9. Family and Kinship Organization. The largest kin group among the peons of the Laguna haciendas was the joint family occupying a single household. Commonly, the family heads in

these households were father and son or brother and brother; occasionally, newly married couples lived with the bride's parents. In general, joint households were not maintained for long periods because economic necessity required periodic migration. The ties between husband and wife do not seem to have been strong, judging from the large number of separations. Ritual kinship was extensive but the bonds did not have great importance in terms of rights and duties. In the traditional hacienda, migration was impossible for the resident peons, so the communities inevitably became closely related. Our data on family organization in these haciendas are scattered, but it can be assumed that if joint households were maintained, they were more stable than in the Laguna region, where migration was common.

10. Religion. Few Laguna haciendas had a church, and most of the people attended church in the towns infrequently. Baptism was the only religious rite that was universally observed. The Catholic Church exercised little or no influence in political or economic matters, and the hacienda owners made little attempt to bring pressures to bear through Church channels. Most classic haciendas had a church; the resident priest was supported largely by the hacienda owner and used his religious authority to promote obedience to the rules of the existing social system. On the question of church rites, the differences between the two types of hacienda do not seem to have been appreciable.

11. Demography. Because of the comparative lateness of the development of the Laguna region, the inhabitants of most of the haciendas were first- or second-generation immigrants. And the fluctuating water supply and system of nonresident wage labor caused a continued internal migration. The workers in the typical Mexican hacienda came from families that had been in the same locality for many generations; internal migration was difficult, and sometimes impossible, because of debt bondage passed on from father to son.

12. Education. No clear distinction can be made in this connection between the Laguna hacienda and the classic hacienda; neither offered much in the way of formal education to the workers'

children. However, the relative wealth of the Laguna haciendas and their proximity to the urban center of Torreón, with its government officials, may have encouraged compliance with the law regarding the establishment of schools in hacienda communities. San Miguel, at least, had a school soon after 1904, when the hacienda was established.

In short, the Laguna hacienda was more a part of an urban-industrial complex than a self-sufficient plantation, the form most often taken by the Mexican hacienda. Still, it would be misleading to present a picture of homogeneity within the region. Though the Laguna haciendas generally conformed to the pattern presented, they were by no means all of a type with respect to ownership and administration. In this connection, three Laguna haciendas can be distinguished. One was the stock company hacienda, of which the outstanding example was the Tlahualilo Land Company, the largest and most highly capitalized estate in the entire region. Of all the Laguna haciendas, it was probably the least like the traditional hacienda. A second type was represented by the hacienda of San Miguel. This was the leased hacienda, an arrangement in which the landowner, unable to undertake the large capital investments required, leased his estate to a company in return for an annual rent. By and large, the leased hacienda and the stock company hacienda were alike; the chief difference was in the divided ownership of the soil and the capital equipment. Finally, the Laguna region also had owner-administered haciendas, of which the hacienda of La Paz was a notable example. Though still far removed from the traditional, self-sufficient estate, La Paz had some of the same paternalistic features. The owner built better than average dwellings for his workers and created a model community, which is still a showplace in the Laguna region.

Formation of the Ejidos. Though the Mexican Revolution was fought, to a large extent, to reform the system of land tenure, there was no fundamental change in land tenure in the Laguna region until 1936. In the two decades that followed the passage of the 1915 Agrarian Laws, which established the ejido system, only 11 small ejido grants were made in the region. Until 1922 applications for ejido grants could not be made by workers who lived on a hacienda (*peones acasillados*) or by the inhabitants of towns large enough to have the political status of *villa*, except in certain special circumstances. Much of the rural population was thus excluded. Efforts to apply for ejido grants or to organize and strike for better working conditions were largely unsuccessful until 1934, when Cárdenas became president of Mexico.

In the depression of the early 1930's, the world price of cotton fell and the economic situation in the Laguna region became desperate. The hacienda owners responded to the crisis with further mechanization in an attempt to reduce labor costs, to which the workers reacted by renewing their efforts to organize unions and obtain ejido grants. In an effort to silence the demands of the new unions, the landowners of the region, acting through their Chamber of Agriculture, offered to buy land for the creation of ejidos if the president of Mexico would then declare agrarian reform achieved in the region and give them written assurances that their lands would never again be subject to expropriation for ejidos.

In 1930, when this agreement was accepted by the Ministry of Agriculture, the 11 ejidos already in the Laguna region had 2,318 members and roughly 5,000 hectares of irrigable land. With one exception, these ejidos were on the periphery of the region, with poor access to irrigation water. The new ejido grants provided some 5,300 hectars to 1,025 heads of families; but instead of solving the problem this action aggravated it, not only because the land was of poor quality with insufficient water but because the number of workers who were granted land represented only 3 per cent of the rural workers in the region. The workers claimed that intimidation was used when the census was taken to determine the number of eligible workers, and that preference was given to union organizers and other "agitators" in an effort to get them away from the haciendas (Senior 1940: 145).

The election of Lázaro Cárdenas to the presidency gave impetus to the unionization of the rural workers and to the demand for ejidos. In a pre-election speech at Torreón, Cárdenas voiced his approval of labor unions, urging the rural workers to organize and demand their constitutional rights. The response was immediate, and in 1935 there were 104 strikes by rural workers in the region. The first strike, which took place on Hacienda Manila, was for a collective contract, a daily wage of 1.50 pesos, an eight-hour day, and the right to name a checker when cotton was weighed. Strikers at other haciendas made similar demands. The landowners responded with the wholesale firing of union organizers. In May 1936 urban labor unions joined the rural unions in a general strike to protest these firings, which prompted the government to appoint an investigating committee to study the situation. But when findings favorable to the workers were announced, the employers refused to accept the committee's recommendations. Another general strike was then called for August 18, with the avowed goal of obtaining a single collective contract for the roughly 28,000 rural workers in the region. Declaring the strike illegal, the State Labor Services of Coahuila and Durango sought and obtained troops to protect 10,000 strikebreakers, who were brought in from outside the region. At this point, Cárdenas intervened, calling the strike leaders to Mexico City, requesting them to call off the strike, and promising to apply the Agrarian Laws in the region. Soon after, on October 6, he signed a decree creating collective ejidos, which contained the following provisions:

1. Twenty or more persons must join in applying for an allotment of ejido land.
2. The area given to the ejido will be four hectares per ejidatario (or slightly more if the land is not well watered or is of poor quality).
3. The property to be affected must be within seven kilometers of the residence of the workers who apply for it.
4. Persons applying for a grant must have worked six months in the region.
5. Those eligible to apply must be males over sixteen, or married, and must be capable of working the land given them.
6. One hundred hectares of each private property will be inviolate of further expropriation, but if there is no request for it, the owner may keep up to 150 hectares.
7. The location of the land to be retained by the owner will be determined by him.

8. Any land remaining after expropriation beyond the 150 hectares kept by the landowner must be sold in lots of not more than 150 hectares each.
9. Owners will be reimbursed for any expenses incurred in helping to create the ejidos.
10. The land will be paid for in government bonds, the amount being the assessed value plus 10 per cent.
11. Credit will be furnished to the ejidos by the Ejido Bank and to the private property owners by the Agricultural Bank.

(Liga 1940: 44–55)

The collective form of grant was preferred over individual ownership and operation because of the obvious benefits in efficiency in large-scale operations. Furthermore, the Laguna workers were accustomed to this kind of operation, only now they would work under the direction of an elected work chief. Though the welfare of the workers was important to the government, so was the maintenance of a high level of agricultural production. Such a level, it was felt, could be maintained only with large-scale operations.

Other economic advantages were claimed for collectivization as well: it lent itself to the use of expensive farm machinery, provided for the use of persons with vocational specialties, such as carpenters, mechanics, and tractor drivers, made the rotation of crops more feasible, permitted the development of off-season enterprises, so as to make fuller use of labor resources, ensured a more uniform quality of product, and increased the workers' bargaining power in marketing the crop, since they would be in a position to withhold products from the market for a time. A further argument was that various types of social services and provision for widows and the disabled could best be handled on a collective basis. Finally, collectivization would preserve the economic unity of the previously existing hacienda instead of breaking it up into small plots. (Whetten 1948: 211–12.)

The distribution of land and the creation of most of the ejidos were accomplished in the 45 days following Cárdenas's decree. The lack of planning preceding the expropriation and the mechanical application of a formula for land redistribution led to several problems that have threatened the success of the ejidos ever since.

History and Geography

Hasty surveying and a lack of information on the water supply led to the distribution of land that was irrigable only in exceptional years.

The government census of workers in the region eligible to become ejidatarios listed 40,208 adult males, of whom 38,101 were actually given ejido grants. It has been estimated that of the 40,000 eligible workers between 15,000 and 16,000 lived in the haciendas permanently, 10,000 lived in the region permanently (but not in the hacienda communities), 5,000 were migrants who came to the Laguna region every year during the cotton picking, and 10,000 were men who had been brought to the region to break the anticipated strike. (Liga 1940: 57, 109.)

In the first year of the collective ejidos, only 30,452 of the 38,101 ejidatarios were members of ejido credit societies receiving loans from the Ejido Bank; by 1939 the figure had risen only slightly, to 30,519. There were thus some 8,000 ejidatarios who never became a part of a collective ejido.

With the creation of the collective ejidos, the number of permanent resident workers in the agriculture of the Laguna region was twice as great as it had been previously (30,000 as opposed to 15,000); and the amount of income-producing land they had obtained was only two-thirds of the total irrigated area. Consequently, the ratio of workers to land area in the ejidos was three times as great as it had been in the haciendas. In addition, the private landowners were permitted to keep the best land and most of their capital assets—wells, roads, buildings, animals, and machinery. Moreover, the government formula for land distribution left many workers still landless simply because they did not live within seven kilometers of land that could be expropriated; at the same time, many landowners sold thousands of hectares of unexpropriated land because there were no workers living nearby who were eligible for ejido grants. Whetten (1948: 222) suggests that a more efficient method, and one that might have been more acceptable to the landowners, would have been to expropriate haciendas intact.

The creation of the ejidos left the Laguna region divided into

TABLE 2
Distribution of Laguna Land

Owner	Total land		Irrigable land		Number of land units	Number of landowners
	Hectares	Per cent	Hectares	Per cent		
Ejidos	405,030	81%	149,139	71%	336	30,337
Private owners	93,128	19%	61,591	29%	2,394	2,394

SOURCE: Secretaría de Recursos Hidráulicos 1951: 221–23. The source does not make clear exactly what year these data are for. The period discussed is 1944–48.

two economic sectors: the collective ejidos, with over two-thirds of the irrigable land, and the remnants of the haciendas, broken up into private properties of 150 hectares or less.

Table 2 indicates that the ejidos are the dominant form of land tenure in the Laguna region, but the important factor is not the amount of potentially irrigable land but the amount of land actually irrigated. The private landowners own most of the wells in the region; moreover, since 1940 they have received a highly favorable distribution of Nazas River water. Consequently, in some years they have cultivated more land than the ejidos. In any case, it is impossible to determine the true situation of land ownership among the pequeños: many apparently independent properties are in fact only held in the names of various close relatives and remain part of a single large enterprise, a legal fiction maintained

TABLE 3
Irrigable Land of Private Farms by Size of Unit

Size of unit (*hectares*)	Total area (*hectares*)	Number of owners
0– 5.0	1,482	691
5.1– 10.0	4,014	495
10.1– 20.0	8,414	486
20.1– 30.0	2,745	108
30.1– 40.0	3,803	115
40.1– 50.0	7,083	157
50.1–100.0	17,527	225
100.1–150.0	16,519	117
Total	61,591	2,394

SOURCE: Secretaría de Recursos Hidráulicos 1951: 224. The source does not make clear exactly what year these data are for. The period discussed is 1944–48.

only to comply with the law that one person can own no more than 150 hectares. Table 3 shows that though the median-sized private farm is 10 to 20 hectares, more than half the private properties are 50 hectares or more.

In all about 300 collective ejidos were created in the Laguna region in 1936 and 1937. In the ensuing decade some were subdivided, so that by 1948 there were 361 ejidos, of which 275 were receiving Ejido Bank loans. Table 4 shows the distribution of land, population, irrigable area, and other property among all 361 of them in 1948.

Before turning to an examination of the effect of the change in land tenure on the ejido communities as reflected in the experience of San Miguel, a word about the effect of that change on the private properties. With the forced sale of land and capital goods, the private landowners had more capital to use on considerably less land. One effect of this was to give them an opportunity to experiment with dairying and with diversified agricultural production; another was to facilitate investments in commerce and industry. At the same time, relationships with their workers

TABLE 4
Characteristics of 361 Laguna Ejidos, 1948

Category	Median	Range of the median 67 per cent	Smallest	Largest
Number of ejidatarios	78	39–127	15	365
Total irrigable land (hectares)	350	200–600	70	2,236
Hectares per ejidatario	4.8	4.0–6.5	1.9	19.2
Hectares of cotton planted	100	50–200	10	600
Cotton yields (metric tons per hectare)	1.0	0.9–1.5	0.5	1.9
Wheat yields (metric tons per hectare)	1.2	1.0–1.5	0.4	2.2
Debts (in thousands of pesos)	42	0–175	0	772
Number of tractors	1	0–3	0	9
Number of trucks	1	0–1	0	4
Number of mules	35	15–82	5	245
Number of wells	1	0–3	0	8
Per cent of loans repaid	96%	80–100%	0	100%

SOURCE: Ejido Bank, Torreón, 1953.
NOTE: Comprises 275 ejidos operating with Ejido Bank credit and 86 without.

became even more impersonal. Fearing to let the workers live on the land, the owners hired only day laborers. In short, the net result of the breakup of the large estates, as far as the private farms were concerned, was an increase in the capitalization of agriculture, increased profits per hectare, and the transformation of private properties from a partly traditional, paternalistic form of social organization to a more fully capitalist form, in which relations between worker and employer were strictly on a cash basis.

2
Description, Demography, and Formal Organization

Physical Description

Flat and straight, the smooth new highway streaks westward from Saltillo, capital of Coahuila, across 200 miles of desert sand until it reaches the oasis of the Laguna region at the base of the Sierra Madres. Abruptly the windblown sand and sparse desert vegetation are transformed into geometric fields of green, whose clean, straight lines reveal the limits of irrigation, the boundary between life and death, between living, verdant cropland and the dry, gray barrenness of the desert.

Scarcely a mile past the first green fields of cotton, the highway comes to Matamoros, oldest town in the region and renowned for the successful struggle of its free farmers against a powerful landowner who, less than a century ago, claimed the soil and water of the entire Laguna region as his private estate. On the main street of Matamoros, a rough interruption of the smooth, modern highway, stands the large Catholic church, and next to it the market place. More frequently visited by the people of the surrounding ejidos, though, were the Ejido Bank on the plaza and the branch clinic of the Ejido Medical Service, whose hospital and headquarters are in Torreón. Here a few doctors and nurses have been struggling since 1936 to provide basic health care for the ejido communities of the entire municipio of Matamoros.

Four miles west of the town of Matamoros, the glistening metal water tower and tin-roofed cotton gin of the ejido of San Miguel rise as abruptly from the desert plain as the first ranges of the

Sierra Madres, ten miles to the west. Two miles south of San Miguel, but hidden from the highway, the dry river bed of the Aguanaval River meanders toward Matamoros and its ultimate destination, a dusty depression that was once a shallow lake, the Laguna de Viesca. An equal distance to the north, the Nazas River curves from Torreón through the heart of the Laguna region toward Lake Mayran and the town of San Pedro. Located midway between the Nazas and the Aguanaval, San Miguel was assured of water for its fields when either river flowed, for its canals connected with both.

Paralleling the smooth east-west highway is the railroad track from Monterrey and Saltillo to Torreón. But the railroad and its small daily train are fading symbols of the era when foreign-owned plantations like the Tlahualilo Land Company dominated the economy of the region. The highway has become the primary link between the ejidos and the rest of Mexico, the means by which most of the people and goods of the region are transported and the symbol of the modern era of economic interdependence. Over the highway the trucks of San Miguel traveled daily to Torreón and Matamoros for gasoline, diesel oil, farm equipment, mail, and other supplies; at harvest time they carried the ejido's cotton and wheat to market in Torreón. Every 20 minutes the bus between Torreón and Matamoros stopped at the wooden shelter in front of the school, picking up men on their way to work, women going to market, students headed for high school in the city; it brought to San Miguel merchants, relatives from other ejidos or from the city, government officials, nurses of the Ejido Medical Service, and other members of the larger society.

Six miles west of San Miguel, the highway reaches its terminus, the modern industrial city of Torreón, hub of the Laguna region and a major economic center for both Coahuila and Durango. Here and there along the road from San Miguel to Torreón, rows of great green piñavete and álamo trees stand astride the canals and dot the roadside. At night the electric lights of the ejidos and of each of the innumerable wells could be seen in every direction. The sense of proximity to many communities is strong, a feeling of closeness that, for the inhabitants of San Miguel, was reinforced by kinship ties with people in the surrounding villages.

The ejido of San Miguel was a long rectangle, 1.4 kilometers wide and 7 kilometers long, which was divided by the highway into two unequal parts: the southern third contained the village and the corn fields, the northern part most of the irrigated cropland. Across the highway the men of the ejido went to the fields in the early morning, some on horseback or on jogging burros, others on foot behind a mule-drawn plow. All were agricultural workers, but they had very different economic roles and statuses. Some owned the land they were going to work, others were landless laborers (*peones libres* or simply *libres*) on their way to perform an obligation owed their kinsmen, still others were on their way to work collectively under the direction of an elected work chief.

Most of the men were riding as they crossed the road, for it was almost five kilometers to the other end of the village lands, where a faint track marked the boundary with the adjoining ejido of Santa Fé. To the east and to the west, irrigation canals formed boundary lines with other ejidos and private estates, but to the south there was no visible boundary, for here the lands of San Miguel merged with barren, unwatered desert.

Along the front of the ejido, facing the road, were the cotton gin, the schoolhouse, and the baseball field, three visible symbols of the economic and social change since 1936. The baseball field was the scene of weekly games between the ejido team and teams from other ejidos or from Torreón; it was also used several afternoons a week for practice. The bleachers of rough-hewn wood, dark and unpainted, provided seats for more than 100 spectators and a protective shade from the fierce afternoon sun.

Next to the baseball field, and occupying the central position at the front of the ejido, was the schoolhouse, a point of community pride, as indicated by the fresh paint, the well-watered bushes, flowers, and young piñavete trees, the high, strong, wire fence to protect the children, the elegant outdoor drinking fountain. The school had three rooms, three teachers, and four grades,

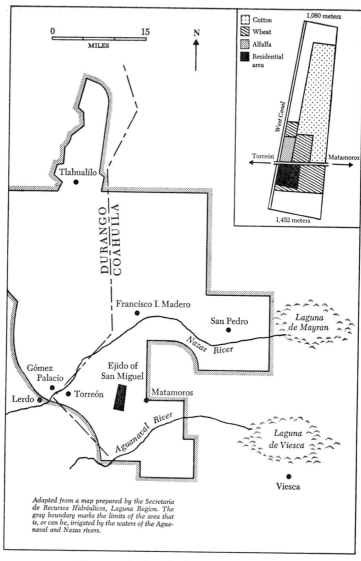

Laguna Region, Coahuila and Durango

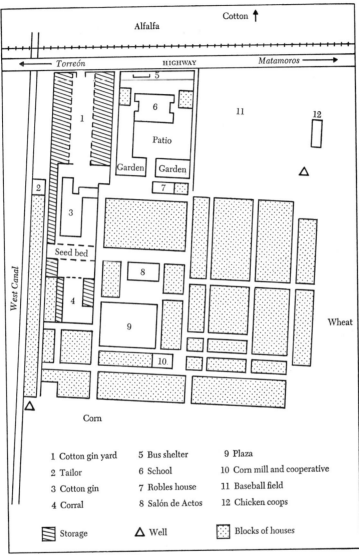

Residential Area, Ejido of San Miguel (1970)

and plans for expansion were being made. Education was clearly valued in the ejido, and the school one of the most important improvements since the ejidatarios took charge.

Beside the school but farther back from the highway, the cotton gin dominated the village skyline, just as cotton dominated the economic life of the ejido. Built in 1948 with ejido savings, the gin was testimony not only to the community's past successes but to its ambitions for further progress through industrialization.

Behind the school and the cotton gin, more than 150 adobe houses formed a rectangle, a pattern that developed in the hacienda period when the dwellings were arranged in the shape of a "U" facing the railroad. The central area was largely vacant in that period, enclosing only the barn, the corral, and the large white casa grande, home of the land company administrators. After 1936 the single row of small, adobe huts on each side was replaced by two to four rows of larger, better constructed houses, and the open central area was almost completely filled with small stores, an assembly hall, a theater, a barbershop, two pool rooms, and an uncompleted church. Part of the central area remained vacant and was occasionally used for fiestas, but it was not a plaza in the traditional, formal sense. The social activities usually associated with the plaza in Mexican towns and villages—promenading, dancing and street music, official ceremonies, lounging about—were carried on in other parts of the ejido; the central area was used only when a particularly large space was required.

On the northern side of the open center, and facing away from it, stood the long, white ejido assembly hall with its red tile floor and rows of hard wooden benches. Here the ejidatarios met on the first night of every month to discuss and decide important community matters. And here, too, the libres' Agrarian Committee gathered every other week to plan the effort to achieve ejidatario status. Directly opposite the assembly hall, on the south side of the open central area, was the unfinished adobe shell of the church. Begun several years before 1953 and deteriorating almost as fast as it was being built, the church suggested that organized religion was not very important to the people of San Miguel.

Demography and Formal Organization 31

One block east of the church the business section of San Miguel faced a wide, dusty thoroughfare that led from the baseball field and the highway on the north to the government-sponsored consumers' cooperative store at the southern end of the ejido. Most of the small unpainted adobe stores lining the street—the barbershop, the pool halls, the butcher shop, and two small stores—adjoined or were part of a dwelling, but the blue, high-roofed cooperative with its brick porch stood separate and apart. Here every morning the women came to grind their corn at the cooperative mill, and in the afternoons several men always lounged nearby, discussing matters of interest—the weather, the crops, the baseball team, the high prices and poor quality of beans and corn.

Most of the dwellings of the ejido were unpainted adobe structures of two rooms, with thatched roofs, dirt floors, and adjacent yards enclosed by a high adobe wall, to keep animals in and strangers out. Each ejidatario was allotted a house site 20 meters square, but brothers or fathers and sons often combined their lots and lived together. Flowers and young trees added a touch of color to every yard, and the doorways of some houses were almost hidden by a profusion of green.

Though all the houses built after 1936 were larger and better lighted, ventilated, and furnished than the shacks built by the hacienda owner, there seemed to be no fundamental architectural changes in most of them. Some, however, were clearly different. These houses, of a superior construction, belonged primarily to ejido leaders. Usually brightly painted—pink, green, blue, yellow, or white—or sometimes of brick rather than adobe, they had tile or brick floors and tile roofs; some even had running water, metal window screens, and a chimney for the kitchen stove. These houses were within the financial means of most of the ejidatarios; primarily all that was required was a desire and an ability to budget one's income (a trait that unfortunately was not developed in the earlier hacienda period).

In striking contrast to these cheerful cottages were the small, dirt-floored hacienda dwellings, which were still occupied by some families. The few ejidatarios who lived in these hovels either were very old or were disposed to spend their income for things other than a better dwelling. The rest were occupied by libres, most of whom could not afford a house and found the old hacienda dwellings the only places available to them, unless they were able to share a house with an ejidatario relative. The sharp differences in the three types of dwellings in San Miguel clearly reflected two fundamental distinctions within the community: the economic difference between the ejidatarios and the libres, and the sociopsychological difference between the ejidatarios who had embraced urban attitudes and those who had retained more of the attitudes developed as peons during the hacienda era.

Demography

At the end of the hacienda period in 1935, the population of San Miguel was almost the same as it was two and a half decades earlier. As an ejido, however, its population has increased twice as fast as the population of the state of Coahuila, of which it is a part; between 1935 and 1952 it grew at a phenomenal rate, doubling its population in 14 years.

Table 5 shows the population of San Miguel from 1900 to 1960, as recorded in federal censuses. Of the 187 males recorded in the 1910 census, an estimated 108 were adult male workers. By 1921, at the end of the Revolution, the male population had been reduced

TABLE 5
Population of San Miguel, 1900–1960

Year	Males	Females	Total	Per cent increase during decade	
				San Miguel	Coahuila
1900	1	1	2		
1910	187	158	345		
1921	139	145	284	−18%	−19%
1930	174	155	329	16	11
1935[a]	245	219	464		
1940	261	241	502	53	26
1950	459	420	879	75	31
1952[b]	534	538	1,072		
1960	551	514	1,065	21	26

[a] Agrarian Department census.
[b] Ejido Medical Service census.

to 139; most of the 48 who left were probably adult male workers.* During the 1920's San Miguel's population increased, and at the end of the decade it was close to the pre-Revolution figures—329 in 1930 against 345 in 1910.

In the decade of the 1930's, during which the ejido was created, San Miguel's population increased 53 per cent, and during the 1940's the community grew even more rapidly. Comparing San Miguel's growth with that of the state of Coahuila shows that they had a similar rate of growth during the two hacienda decades, from 1910 to 1930, but that between the creation of the ejido and 1950 the growth rate of San Miguel was more than double that of Coahuila.

In addition to the government's ten-year censuses, a special census was taken by the Agrarian Department in 1935 to determine the number of adult males in the community who were eligible to become ejidatarios; and in 1952 the Ejido Medical Service made a census for its own use. Both censuses show higher totals than those arrived at by interpolation from the regular government census, but these differences can be explained in one case and dismissed as negligible in the other. On the basis of the government census figures for 1930 and 1940, the estimated population in 1935 was 416, which was 48 fewer than the actual census figure of 464. There is reason to believe that a number of migrants settled in San Miguel in 1935 just before the census was taken, so that the discrepancy between the 1935 census and the estimate probably reflects an actual population increase in 1935 rather than a census error. The Medical Service census of 1952 listed 1,072 names. An estimate of the 1952 population based on the 1950 government figure of 879 and the 7.5 per cent annual increase in the 1940–50 decade yields 1,016, a discrepancy of less than 5 per cent.

The Medical Service census listed 277 adult males—124 ejidatarios and 153 non-ejidatarios. Two other lists of non-ejidatarios, compiled by the Agrarian Committee of San Miguel and by the

---

* These estimates are based on the assumption that there were equal numbers of females above and below fifteen years of age (as in 1953) and equal proportions of boys and girls below sixteen (as in 1953).

San Miguel paymaster (*rayador*), included 54 men whose names did not appear in the Medical Service census. Most of these men were probably not residents of San Miguel but were simply relatives, friends, or migrant workers who occasionally visited or worked there. Since the purpose of the Agrarian Committee was to organize the libres and to exert political pressure to obtain land for them, it is reasonable to suppose that any libre, no matter how temporary his stay in San Miguel, would want his name on the committee's list.

A point to be emphasized in later descriptions of the changing economy and the social organization of San Miguel is that the ejidatario population of the community has remained relatively stable since 1935; the increasing population consists largely of the children and grandchildren of the ejidatarios (plus some migrants), only a small number of whom can expect to become ejidatarios through inheritance.

Birth Rate. The annual birth rate, estimated from the number of children under one year of age in November 1952 and, again, on the basis of the number of known pregnancies in May 1953, was between 54 and 62 per thousand total population. Most women in San Miguel began bearing children at about age seventeen and continued to do so until they were in their mid-forties. In 1952 married women between the ages of sixteen and thirty-one had 611 babies per thousand, and those between the ages of thirty-two and forty-six had 245 per thousand, a rate that indicated the average married woman had 11.8 children during her lifetime. On the average, women between seventeen and thirty-one years of age had a baby every 19 months. (My estimate did not take into account miscarriages or children who died between birth and one year of age.)

As far as could be determined, birth control was practiced only rarely, and then only by continence. The men in the community appeared to have little interest in limiting the size of their families, in spite of the land shortage and the lack of economic opportunity for their children. From the standpoint of the ejidatario, economic conditions were better than they were in the past. Chil-

Demography and Formal Organization   35

dren were not considered an economic liability, and in fact could even be of some help in working the land. In any event, the expenses of raising a child were minimal. Children did not have to be sent to school beyond the free elementary grades, and they married and began to support themselves at about the age of sixteen.

Mortality Rate. Of the 143 male ejidatarios who were participating members of the ejido in its first year, 27 died between 1936 and 1952, an annual mortality rate of roughly 1.3 per cent. No data on infant mortality were obtained, but it seems probable that the rate had decreased as a result of the Ejido Medical Service's work. Before 1936 no medical services were provided by the landowners; nor could the peons afford private doctors. Once the Ejido Medical Service established a hospital in Torreón, many ejidatarios availed themselves of its professional services, and half of San Miguel's babies had been born there. Libres, as well as ejidatarios who did not pay the yearly fee of 150 pesos, could not use the hospital.) The medical service also provided visiting nurses to teach the women of the ejido elementary principles of infant hygiene and nutrition. Nearly all children were vaccinated for the common infectious diseases of childhood. Since none of these medical services were available before 1936, infant and child mortality had almost certainly decreased since that time. However, intestinal diseases, the principal cause of death among Mexican children, were still a scourge in San Miguel. In spite of its best efforts, the Ejido Medical Service had not been successful in persuading all of the mothers to boil water for their babies, to feed them only well-cooked food, and to keep them as clean as possible.

Immigration. After 1936 the right to live in San Miguel was determined by the ejidatarios rather than by the labor needs of the land company. The possibility of living with a kinsman who owned a house in the community and the fact that some work was available to non-ejidatarios attracted many relatives of ejidatarios to San Miguel. In addition, though a number of the ejidatarios' children married outside the community, many brought their spouses to live in San Miguel. According to the census figures, almost 30 per cent of the ejido's adults (171 of 577 persons) had migrated to San Miguel between 1936 and 1952. Of these, 69 were already married and 77 (55 of them women) married San Miguel residents. Eighty-one of the children in the community had migrated with their parents, and 181 had one migrant parent. This meant that 48 per cent of the children had at least one parent who had come to San Miguel after the ejido was established.

Meanwhile, at least 186 persons left San Miguel or died in the same period (exclusive of all those born after 1936). In 79 cases it was not clear whether the person died or emigrated; assuming the same death rate for these 186 as for the ejidatarios who remained, the total number of emigrants was 151. Since there is no way of determining the age of each emigrant when he left San Miguel, it is difficult to compare the figures on migration to and from the community. If the age distribution of the emigrants was about the same as the age distribution of the rest of the community, the net in-migration of adults from 1936 to 1952 was in the neighborhood of 95, or one-third of the total adult population's net increase during the ejido period.

Almost all of the migrants who came to San Miguel after 1936 were from neighboring communities. Some young men who emigrated temporarily to find seasonal employment were accompanied by wives from other parts of Mexico, but on the whole the immigrants in the hacienda days had come from more distant areas than the post-1936 immigrants.

Formal Organization

The Sindicato. The forerunner of the organizational structure of the ejido of San Miguel was the *sindicato*, or labor union, established in 1934. All earlier attempts to organize the rural workers of the Laguna region either had failed or had been short-lived because of the opposition and influence of the landowners. After Cárdenas became president, however, the government began openly encouraging (if not directing) the formation of unions, and within a year workers in every hacienda in the region were organized.

At the same time the government organized an army reserve unit in San Miguel, providing its members with rifles and weekly drill instruction. Thereafter, police duties in the community were performed by the local reservists, a fact that greatly strengthened the bargaining power of the sindicato: the landowners were deterred from meeting the workers' demands with guns by the presence of reservists armed both with rifles and the knowledge that the government was on their side.

Nevertheless, the growth of the San Miguel sindicato was slow, even after it had become clear which way the government's sympathies lay. Fear of loss of employment (and with it the right to remain in the community) kept many peons from joining, and loyalty to the land company and the hacienda owner stopped others, especially the mayordomos. A year after its formation, the San Miguel sindicato had only 38 members, one-fourth of the hacienda's workers. Interestingly, the leaders in this period were not the men who were to hold the most important elected positions in the ejido later; in general the early leaders seem to have been men who were most easily impelled to action and who acted more as individuals than as representatives of extended families or large segments of the community.

The first major success of the sindicato came in November 1935, when the land company agreed to recognize the union and to sign a collective contract. The membership at the time was only 44, but within two weeks there were 105 members; almost every worker in the hacienda joined. The contract that was eventually signed not only recognized the sindicato and its power to bargain for wages and working hours, but also guaranteed the workers an adequate water supply and better maintenance of their houses. The land company agreed also to arrange credit so the sindicato could purchase a mill to grind corn.

San Miguel did not achieve this contract because of the workers' solidarity and willingness to engage in a collective struggle; nor was the contract the result of a sudden benevolence on the part of the land company and the hacienda owner. Rather, the moving force in the situation was the attitude of the federal government, an attitude that was reflected at the state and local levels as well. The absence of strong grassroots support for the union until success was assured by the government is clear in the developments in the sindicato's two-year history. There was a rapid turnover among the officials and in the end the resignation of the secretary-general and one-third of the membership, who formed a rival company union. This division not only indicated the power of the land company, but also revealed the lack of solidarity among the workers, even in the face of a common external threat and in spite of the promise of immense rewards through collective action. Even after the sindicato was successful in gaining recognition as a bargaining agent, it struck only once, and then only to participate in the general strike of 1936, which was not an action against the hacienda of San Miguel but an action against the hacienda system of the entire Laguna region. Further, the general strike, which led to the presidential order creating the collective ejidos, was directed by labor leaders outside San Miguel. It can hardly be said, then, that the sindicato's success in obtaining a grant of ejido land was the fruit of determined and united effort by the workers. The fact is, every important action of the union had been initiated by groups outside San Miguel, from the formation of the sindicato itself to the final successful petition for an ejido grant.

In the sindicato elections of March 1936 the original officials were replaced by another set of leaders, among whom were the heads of the largest extended families in the community. After the ejido was formed and the sindicato was dissolved, it was this second group of sindicato officials who became the first ejido officials.

Ejido Membership and Inheritance Rights. By President Cárdenas's decree of November 25, 1936, the collective ejido of San Miguel became an economic organization whose members jointly owned and operated an agricultural estate of 1,000 hectares, 60 per cent of which was classified as irrigable. The constitution of the ejido, written by the federal government, explicitly mentioned only economic functions. Nevertheless, the ejido assembly inevitably performed political and social functions as well.

Demography and Formal Organization 39

All resident males over sixteen who could work were eligible to become members of the ejido when it was formed. Apart from a small number of men (a few who were too old to work, two or three who left the community in order to get the larger land grants due them as veterans of the Revolution, and two or three of the mayordomos, who refused to become ejidatarios either because they were opposed to the ejido system or because they had strong personal attachments to their former employers and jobs), every eligible male became an ejidatario; moreover, some boys as young as fourteen claimed to be older and were included in the 1936 census. The total number of men sixteen years of age and over in the 1935 census of San Miguel was 150. On this basis, the ejido was granted four hectares of irrigated land per man, four hectares for the school, and 396 hectares of non-irrigable land. Twenty additional men were elected to membership in the ejido after the original grant, but by 1953 only 131 of the 170 ejidatarios were still ejido members. Thirty-five ejidatarios had left the community voluntarily, three had been expelled from membership, and one had died without an heir.

The right to be an ejidatario could be transferred only by inheritance. By law, the widow of an ejidatario automatically inherited his grant; in most instances the land was then worked by a son, who would ultimately inherit the grant. In the event the widow married another ejidatario, the land reverted immediately to the son. If an ejidatario had no wife, he could leave his rights to anyone he wished, though a son or daughter was the preferred heir. If he had no wife or children and designated no heir, the ejido assembly agreed on the disposition of his land. After 1938 the ejido decided on a policy of dividing such land among the remaining ejidatarios instead of naming a replacement. By this process the individual shares had been increased to four-and-a-half hectares, and the ejidatarios indicated they intended to follow this policy until each member had the full legal maximum of eight hectares.

Most of the rules of inheritance were informal and had developed within the ejido over a period of years. The federal law did not recognize the existence of individual *parcelas* (plots) of land,

and so did not regulate inheritance rights in this regard; it merely specified that a widow should inherit the membership rights of her husband. Very often in San Miguel it was difficult to tell whether a man was a *"remplas"* (substitute), working for his mother or for a sister living in his household, or whether he himself had inherited the ejidatario rights of his deceased father. Nor did official government records clear up the situation, since they had not been kept up to date and continued to include the names of deceased ejidatarios rather than the names of the heirs.

The rights of the 27 ejidatarios who died between 1936 and 1953 were disposed of as follows: 16 passed their rights to their wives, six to sons, two to daughters, one to a nephew, and one to a non-relative; one man left no heirs. Only male heirs had the full rights and duties of ejidatarios, for only men could work in the planting and cultivation, and normally only men attended the ejido assemblies. (Informants told me that women participated in the assemblies when very important issues were being decided, but none did so while I was in the community.) As more and more women inherited ejidatario rights the decision-making of the ejido had been placed in the hands of a smaller and smaller proportion of the total community.

The members of the ejido met at least once and usually two or three times each month to discuss and vote on important matters. Though the legally designated functions of the assembly were primarily economic, it was the de facto civil government of San Miguel, and so made decisions concerning other vital aspects of community life. For example, it approved or rejected immigrants who desired to live in the community; it granted permission to both ejidatarios and libres for use of land for a house site; and it called on resident libres to participate in collective tasks as needed. As the primary non-kinship group with money, it determined community activities by giving or withholding financial support. For instance, it had sponsored and given large amounts of money to such enterprises as the baseball team, the fiestas, and the school; and it had refused to support the building of the church.

Many of those who fought to obtain the collective ejido grants asserted that the economic dominance of the hacienda owners and the land companies resulted in the social and political control of the hacienda community. The consequences seem to have been much the same since the ejido has held economic power: it too controlled the social life of the community, including the libres, who were excluded from full participation. Only the ejidatarios, representing roughly 40 per cent of all the males sixteen years old and over, voted in the election of community officials and participated in major community decisions.

The Ejido Assembly and Officials. The Federal Agrarian Code required all collective ejidos to have a general assembly, which was charged with the following duties and responsibilities: to elect and remove the members of an executive committee and the members of a vigilance committee in accordance with the provisions of the code; to authorize, modify, or rectify the decisions of the executive committee whenever this is in order; to discuss and approve the reports rendered by the executive committee and to order that an approved statement of account be posted in a visible and central place; to request the intervention of the federal authorities on matters relating to the suspension or privation of rights of ejido members; and to issue rulings on how the communal lands of the ejidos should be used, subject to the approval of the Ministry of Agriculture or the National Ejido Bank. (*Nuevo codigo agrario* 1943: Art. 42.) Seemingly this assembly was to have considerable power; in practice, however, the Ministry of Agriculture, the Ejido Bank, and the Agrarian Department have been in a position to veto almost any decision.

The executive committee of San Miguel, made up of a *socio delegado* (member delegate), a secretary, and a treasurer, elected for a term of three years, had several functions. It represented the ejido with power of attorney before administrative and judicial authorities; it administered ejido property that was used collectively; it oversaw the division of collective lands into plots; it called a meeting of the general assembly at least once a month, or whenever the vigilance committee, the Agrarian Department, the Ministry of Agriculture, or the Ejido Bank so requested; it kept the general assembly informed on the work carried out in the ejido, accounted for the expenditure of funds, and advanced proposals as necessary; and it followed and enforced the rulings of the ejido assembly.

The reelection of officials, though permissible, seldom occurred in San Miguel. Election to office was clearly not just an honor and a privilege, judging by the frequent resignations, especially on the part of the less important officeholders. Though the officials received slightly more than the daily wage for collective or parcela work, and were paid for the full seven days in the week, the responsibilities of office were great; the financial reward was not sufficient in itself to compensate for the time and effort required. In part because of this, some of the community's most capable men had been reluctant to accept office; at least one refused reelection to the highest office, that of *socio delegado*. Outright dishonesty, the only way an officeholder could make substantially more money than his fellow ejidatarios, had been conspicuously rare in San Miguel.

In the ejido's first year there were two officials of equally high rank, the *socio delegado* and the *consejo de administración*, with divided responsibilities according to whether matters were external or internal to the ejido. But it soon became clear that no sharp line could be drawn between these matters, and the two positions were combined. Since the *socio delegado* was the ejido's representative with the outside world in all important affairs, and especially with the Ejido Bank, he had to be able to read and write. And since much of his time (the greater part of almost every day) was spent in Torreón or Matamoros on official business, he had to be able to deal successfully with government officials and bureaucrats as well as businessmen and suppliers. It was thus not surprising to find that the *socio delegados* of San Miguel had almost always been elected from among those most accustomed to urban ways and able to hold their own with city businessmen.

Half the *socio delegados* of San Miguel had come from two families that were closely linked by marriage and had been in San Miguel since the beginning of the hacienda period. The Morales and Argumaniz families contained more ejidatarios than any of the other extended families, and had furnished leaders to the community since long before 1936, the head of the Morales family having been a mayordomo for the land company. In general, the *socio delegados* who had not come from these two families had been less successful as community leaders; two had had bitter experiences with finances. The *socio delegado* had to authorize all expenditures and sign all contracts and agreements for the ejido; thus he was equally liable with the treasurer for any shortages. Since no ejido official had ever had formal training in bookkeeping, discrepancies were understandable; and that these had led to charges of corruption was not surprising. The fact that financial scandal had been confined to *socio delegados* who were not members of the leading families seemed to indicate that they were either less careful or, more likely, less able to gain the confidence of the community when carelessness led to apparent shortages.

Though the position of secretary was an important one, it did not involve as much responsibility as the office of *socio delegado*, and most of San Miguel's secretaries had been young men who had not previously held an office. The secretary's most time-consuming task was managing the storehouse, where he had to weigh or count all ingoing and outgoing items—cotton, wheat, alfalfa, seed, fertilizers, and the like—and make the necessary financial transactions and records. Beyond that, he performed the usual duties of a secretary at the ejido assembly, taking the roll and keeping the minutes of the meeting.

The treasurer kept records on all ejido financial matters, both external and internal, made payments authorized by the *socio delegado*, and received and deposited money owed the ejido. The majority of the ejido's financial transactions were made through the Ejido Bank, which had to approve major expenditures, and since all crop loans were obtained from the bank, it exercised almost total control over important financial decisions. Apparently the ejidatarios did not object to this, for they had the option of saving their money and financing the crop themselves; this course had, in fact, been considered and rejected in the assembly.

The treasurer was also the paymaster for both collective and individual work. He had to draw money from the bank weekly to pay each worker's weekly wages (*anticipos*), and it was his responsibility to check each ejidatario's cotton field to see that the work had been done. The ejido paid out cash wages only once a week, but the treasurer was in his office every afternoon to give pay vouchers to those who wanted an advance on their week's wages. These vouchers were redeemable only within San Miguel; anyone needing cash had to exchange a voucher with a friend or with the moneylender, a private entrepreneur who gave 18 pesos for a 20-peso voucher.

Essentially, the ejido's method of payment was the same as the method used in the hacienda period, and the ejidatarios apparently did not consider it worthwhile to get a week ahead so as to be paid in cash at the end of each day. Though they thought the voucher system a bother, they clearly did not regard it as the kind of oppressive evil it has been labeled in descriptions of the hacienda system. (Tannenbaum 1929: 118–19.)

The ejido had been shaken from time to time by financial scandals, owing, apparently, to a careless handling of routine financial matters. One of the most serious of these was in full swing in the summer of 1953; it was claimed that the books of an outgoing administration showed a deficit of over 6,000 pesos, 5 per cent of the total budget during the period the administration had served. As in earlier instances, the problem seemed one of poor bookkeeping rather than dishonesty, though the matter was still under investigation when I left San Miguel. This same problem appeared likely to arise again and again until the ejido hired a bookkeeper or sent one of its young men to business school. Failing that, the efforts of untrained treasurers, no matter how well intentioned, seemed certain to remain a source of trouble and internal dissension. Already it had become difficult to persuade some of the most capable men in the community to take the position of treasurer, largely because they had no desire to assume extra work and responsibility, and then be accused of dishonesty.

In the first years of the ejido, the work chief, who directed all field labor, was one of the most important officials in the community, possibly more important to the success of the new economic organization than the *socio delegado*. At that time his task was so great as to require several assistants. The first man to hold this position in San Miguel, an ejidatario of great prestige, resigned after three years, finding the work performance of the ejidatarios poor by his standards. This experience led him to be among the first to advocate dividing the land into individual plots.

After 1944, when all the land was subdivided, the position of work chief became less crucial. Nevertheless, he continued to be the second most important official in San Miguel: he decided who would work at each collective task, and the total crop yields depended largely on his agricultural skill and ability to direct the men. Still, all major agricultural decisions were made by the ejido assembly, and the work chief's opinion carried only slightly more weight than those of other knowledgeable and highly regarded ejidatarios.

Though it was considered an honor to be elected work chief and though the position paid as much as that of *socio delegado*, the work and responsibility involved kept the post from being sought after; in fact, several work chiefs had resigned within a few months. The work chief in 1953, a very able man whose brother had held the same post several years earlier, was reluctant to fill the position much longer, even though the ejido assembly was well satisfied with his performance.

Whetten is of the opinion that "the work foreman is perhaps the most important person in the entire ejido setup.... But he is selected by majority vote of the ejidatarios and hence is not always the most competent person available for the job. In some cases the ejidatarios may vote for the one with the most pleasing personality or the one with the strongest political influence rather than for the one best prepared technically for the job" (1948: 226). In San Miguel this had not been the case, in part because the job was not so desirable as to provoke a popularity contest and in part because agricultural skill was a primary basis for status in the ejido. Moreover, the question of the work chief's technical skill was of reduced importance, since the ejido assembly made the major agricultural decisions. The question of worker motivation, which was a serious problem for the first work chiefs, became much less an issue after the decollectivization of the land and the development of a libre group that competed for employment in the collective work.

When San Miguel purchased its cotton gin in 1948, the important position of manager (*gerente*) was created in the ejido. This job was seasonal (the gin operated only three or four months of the year) but crucial, not only because the price of cotton depended on the quality of the ginning but also because San Miguel augmented its income by processing the cotton of other ejidos. The manager's tasks were to direct the operations of the plant, to keep the labor and production records, and to report to the ejido assembly. For part of the ginning season the plant operated around the clock, requiring the services of an assistant manager.

In principle all the ejidatarios were supposed to share equally in the work of the cotton gin, but in practice they were allowed to send libres to substitute for them. The effect of this was to exclude from plant work those libres who did not have relatives or close friends needing a substitute. Though the hourly rate of pay at the gin was about the same as the rate for collective agricultural work, an 18-hour shift made plant work the highest-paying job available to the libres in the ejido. (No one in the community was capable of repairing the gin, or even of keeping it in good mechanical condition. Consequently, the ejido had to call on the services of a mechanic from Matamoros, whose daily pay was three times that of the *socio delegado*.)

According to the Agrarian Code, each ejido had to elect a vigilance committee whose function was to uncover corruption and expose incompetence in the work of the major officials. (The code also provided that in the event of a close vote between factions for the office of *socio delegado*, the losing faction could select the president of the vigilance committee. The issue had never come up in San Miguel; elections were close to unanimous, as were most of the decisions of the ejido assembly.) Since the *socio dele-*

Demography and Formal Organization 47

gado handled most of the ejido's important financial transactions, one of the committee's principal tasks was to keep close watch on his operations. I heard stories of *socio delegados* in the Laguna region who had left their ejidos at the end of a term of office as rich men. San Miguel had been fortunate in having a succession of unusually able and honest leaders, however, and the task of its vigilance committee had been correspondingly easy. The three-man committee was unpaid and met only as the need arose, for instance when there seemed to be a shortage in the treasury. Since individual ejidatarios were reluctant to express criticism of the ejido leaders in public, the vigilance committee was a device that permitted necessary criticism to be voiced as a duty of office rather than the voluntary act of one man.

For each of the three major officials and for each of the members of the vigilance committee, substitutes *(suplentes)* were elected to take up the duties in the event of absence, illness, death, or resignation. Though these substitutes were unpaid, election to office was an honor, and the effect of the system was to double the number of prestige positions in the community. Adding to these the *juez* (roughly, justice of the peace), the work chief, the manager of the cotton gin, the 13 army reservists, the school and fiesta officials, and the members of various special commissions, over one-third of the ejidatarios could hold positions of status at any given time. In the course of a decade, virtually every ejidatario of average ability could occupy an official position, and more than likely some would serve in official capacities many times. Further, the suplente system was a political stabilizer, for the community did not have to hold a new election when an official died or resigned; succession was automatic. The one serious weakness of the system was that the substitutes were not always carefully chosen, so that the ejido might on occasion find itself with a leader it would not have elected. This had happened at least once in San Miguel in connection with the office of *socio delegado*.

From time to time certain ad hoc positions were created in the ejido, which, though not provided for in the ejido constitution, were of vital importance in the administrative work. Primarily these involved the many commissions that represented the ejido in its relations with government officials, business firms, and other organizations outside the community. The men elected to these commissions were usually prestigious present or past officials who were experienced in the administration of ejido affairs. The pay was minimal, and indeed sometimes meant a financial sacrifice for the commission member. This was exceptional, however, since the life of a commission was usually no longer than a month or so. The assignments of these commissions were varied. One, for example, had traveled to Mexico City to arrange the purchase of the cotton gin. Another had gone to Saltillo to talk with the governor of Coahuila about a loan to enlarge the school. Others had journeyed to various cities to negotiate the sale of the ejido's crops.

The Ejido Assembly in Action. Before turning to the processes of group decisions in San Miguel, as reflected in the operations of the ejido assembly, it is important to note that the assembly's deliberations continued informally virtually every time two ejidatarios got together in the fields, in the streets, or in their homes. The formal discussions were carried on in two kinds of assembly meetings: regularly scheduled sessions, which met the first day of every month, and special sessions, which were called when there was need for immediate action on a pressing problem. Both were conducted in the same way, the chief distinction being that the special sessions were usually shorter and their agendas less extensive.

On the afternoon of an assembly a man climbed the water tower and sounded a loud gong to remind the ejidatarios of the meeting, a step that was little more than a formality, since most of them were well aware of the meeting and had already discussed the subjects to be raised among themselves. Though some of the ejidatarios started drifting into the assembly hall at the scheduled hour, usually 8:00 P.M., it was generally 9:00 P.M. before the secretary called the roll. There was a five-peso fine for absence from the meeting, but this rule was not enforced; some of the older and highly regarded ex-officials attended only when they had advice to offer on an important subject.

The presiding officials sat at one end of the hall facing the eji-

Demography and Formal Organization 49

datarios, most of whom sat on the rows of hard wooden benches. Those who could not find seats sat on the floor along the walls or lounged in the doorway. After the roll call and the reading of the brief minutes of the previous meeting, the commissions made their reports, which were fully discussed by anyone who wanted to speak. Next, correspondence was read. Typically this included letters from the Agrarian Department concerning the transfer or inheritance of ejidatario rights or communications from business firms. All were thoroughly discussed, after which the meeting was thrown open to general subjects. Usually the *socio delegado* brought up the first few items. Discussion of every matter was full, and varied from rapid-fire participation to long pauses that gave the less verbal ejidatarios an opportunity to speak their minds. Everyone had the right to speak as often and as long as he wished. There were no parliamentary rules to stifle debate or to rush motions through. The only comments that were censured were those made in anger or those aimed at silencing another speaker. (In one session an ejidatario who had been drinking began to talk rather foolishly. Nevertheless, he was permitted to finish, after which some of his relatives took him out of the assembly hall. Another ejidatario who had objected to the drunken man's ramblings was upbraided for trying to cut off free discussion.)

Debate usually continued until a consensus was reached, or else the question was put aside for a decision at a later meeting. Seldom after general agreement had been reached would more than two or three ejidatarios continue to voice objections by voting against the majority. Since much of the real decision-making occurred in the continuing informal discussion of problems during work or leisure hours, the groundwork for the assembly meeting was well laid, and the voting did not represent decision-making so much as it did the granting of formal sanction by the group. The meeting usually lasted until three or four in the morning. Though a few of the ejidatarios left sooner, most stayed on—some sleeping, a few chatting outside, but the majority attentive to the end.

Despite the free discussion, the ejido assembly of San Miguel had major defects as a democratic community government, among them its exclusion of libres and women, its lack of legitimate civil authority and power, and its heavy dependence on the Ejido Bank and the Central Union.

Political Relationships. One of the principal changes that took place with the formation of the ejido was the sociopolitical shift in the relation of the community to the rest of Mexican society. The Laguna hacienda community was not a legal political entity; its de facto government was partly determined by the hacienda owner and partly by the municipio president (one and the same man in the case of San Miguel).

After 1936 San Miguel's political ties were primarily with the federal government. Though the ejido system increased the decision-making power of the people of San Miguel, it at the same time strengthened the influence of the federal authorities over the community, operating through the Ejido Bank, through the school, through the Ejido Medical Service, and most importantly, through the control of ejido land. The federal government made the formal rules that governed the ejido, and it had the ultimate power to decide whether or not a man could be an ejidatario: the ejido could expel members only with the approval of the Agrarian Department. The increase in federal power was in part simply the result of an increase in services over which control could be exercised.

In the first years of the ejido, San Miguel's ties with the federal government were even stronger than they were in 1953. Not only did the Ejido Bank exercise closer control over agricultural operations, but San Miguel was closely bound to the government through the Central Union of Collective Ejidos, which was almost a branch of the Cárdenas administration.

The Central Union was organized and sponsored by the government to deal with the Ejido Bank on behalf of the many collective ejidos of the Laguna region. When Ávila Camacho became president in 1940, however, he opposed the further extension of the collective ejido system, a primary objective of the Central Union. A rival, pro-government union, the National Union of

Rural Workers CNC (which was in fact a branch of the government political party, the PRI), succeeded in wooing away many ejidos, especially those that had severe financial problems and could be easily influenced by the Ejido Bank. By 1952 the CNC claimed to represent about 250 Laguna ejidos and sectors of at least 27 others, while the Central Union retained the support of only about 40 ejidos, plus the sectors of those that were divided in their affiliation. San Miguel was one of the few ejidos in the municipio of Matamoros that still belonged to the Central Union; 42 belonged to the CNC and five were divided between the two.

As originally conceived, the Central Union was to coordinate action between the ejidos and the Ejido Bank and to disseminate information on agricultural methods, health, education, credit, and related matters. But after 1940 the bank became essentially a lending agency and little concerned with social conditions in the ejidos. The two bodies, with very different goals, became politically opposed to one another, and as a consequence the only area in which the union remained active was the presentation of grievances to the bank, the government, and the newspapers.

The Central Union, which developed out of the advisory committees that operated from 1936 to 1940 as a liaison between the Ejido Bank and the ejidos, was originally intended to be an assembly of ejidatarios, with each ejido sending delegates and with union officials elected from among the delegates. This plan was never carried out: the ejidos did not send delegates and the leadership had not changed since the union's founding in 1940. San Miguel and the other ejidos had little or no influence in the policies or selection of personnel; they could do no more than express approval or disapproval of the union by maintaining their affiliation or resigning.

Nevertheless, San Miguel strongly supported the Central Union and its leader, Arturo Orona, whose opinion on economic and political questions was almost as highly regarded as that of Cárdenas. The union had strengthened its standing with the ejido by successfully presenting a major grievance to the Supreme Court of Mexico. The case involved the ejido's purchase of the cotton gin from an American company in 1948, a transaction that required the Ejido Bank's transfer of money from San Miguel's Social Fund (*fondo social*), in which 5 per cent of the community's gross income was set aside annually for community improvements. As this purchase was being completed, the peso was devalued from a rate of approximately six pesos to the dollar (U.S.) to about eight pesos. The Ejido Bank collected the money from San Miguel at the higher value, then paid the American company at the lower; the difference, pocketed by a bank official, amounted to about 40,000 pesos. The ejido was unable to recover this sum during the administration of Miguel Alemán, but after Adolfo Ruiz Cortínes took office the Central Union presented the case to the Supreme Court and was able to collect the full amount. In return, San Miguel gave 5,000 pesos to a national Communist newspaper, *La Voz de Mexico*, with which the Central Union was politically allied.

The community organization of San Miguel was not closely integrated with the governments of the state of Coahuila or the municipio of Matamoros. The main connection the ejido had with the state was in the payment of taxes. Occasionally a commission was sent to the governor with a petition, as for example in 1953, when the ejido tried to get the governor to intercede with the Ejido Bank, which had refused a loan for a new schoolroom. The ejido's connection with the municipio government was even more tenuous. The activities of the respective principal officers were not coordinated, and though the adult males of San Miguel could vote for municipio officials, they took little interest in the elections.

The only civil official of the municipio who lived in San Miguel was the juez, an office considered so unimportant in the hacienda period that it was held by one of the resident peons. After 1936 the ejido selected the juez, but apart from that, the situation did not change appreciably. The duties of the juez remained trivial; his chief responsibility was to grant permits to peddlers who wished to sell their goods or services in San Miguel; his only payment was the small fee charged for those permits. The juez also supervised the policemen who were assigned from time to time to keep order at dances and fiestas. He was aided in these tasks by an assistant, whom he appointed. The ejidatario who was juez in 1953 had been in office since 1944. A former *socio delegado* and one of the most highly regarded men in the community, he was never referred to by his title and had apparently accepted the position because it required little responsibility. His assistant was a libre of low status.

The people of San Miguel were aware not only of their strong ties to the federal government but also of the effect on their lives of events all over the world. Though their conception of the world economic and political situation was not fully articulated, they were surprisingly sophisticated about such subjects as the Cold War struggle (thanks in large part to the propaganda of the Central Union), the world market prices of cotton and wheat, and the increasing interdependence of San Miguel with other parts of Mexico and the world.

3
Economic Organization

Property

The basic economic change that occurred in the Laguna region when the collective ejidos were created was a change in the system of productive property relations. What was involved was not only a transfer of property rights from one group to another, but, more fundamentally, a change in the very nature of those rights. All of the principal types of productive property—soil, water, irrigation systems, machinery, and animals—were affected.

Water. The distribution of the water of the Nazas River, regulated by federal law and a government water commission from 1890 on, was changed by Cárdenas, who issued a decree in 1936 giving priority to domestic users, ejidos, private properties of less than 50 acres, and large private estates, in that order. However, as interpreted, the new regulations gave the ejidos and the private landowners an amount of water proportional to their areas; under this interpretation the ejidos received two-thirds of the Nazas waters. The distribution of Aguanaval River water was regulated by a water board, to which San Miguel sent a representative. The arrangements concerning water from both rivers had to be worked out every year by representatives of the ejidos, the private landowners, and the national government. In 1953, a year in which the available supply was extremely low, San Miguel received no water from the Nazas, since the ejido's six wells supplied more water than most ejidos got from the river.

Wells continued to be privately owned after 1936, as in the

Economic Organization

hacienda period, but they became subject to governmental control in that the government's permission was required to sink new ones. The two ejido unions in the Laguna region had repeatedly but unsuccessfully petitioned the federal government to regulate the distribution of well water, under the constitutional definition of subsurface water as a national resource. Behind this demand was the fact that the private landowners of Laguna had far more wells in proportion to their acreage than the ejidos. The private owners not only had retained most of the wells in 1936, but had installed more new ones than the ejidos thereafter, since the Ejido Bank had been extremely reluctant to grant the ejidos loans for this purpose. In the drought years between 1950 and 1953, the private farms relied exclusively on well water; at the same time, the ejidos cultivated only two-thirds of their area with subsurface water. The private control of most of the region's subsurface water accounted in no small way for the relative economic success of the private farms as compared with the ejidos.

When San Miguel became an ejido, the hacienda's six well pumps were purchased from the landowner. Many ejidos were not so fortunate, for the landowners usually kept the 150 hectares of land that contained most of the wells. In the case of San Miguel, the owner had so much good land and so many wells elsewhere that he was willing to sell the hacienda in its entirety. Ultimate ownership of the water from the San Miguel wells belonged to the ejido; no arrangements for its use had to be made with a government commission or the local water board.

Water in the main irrigation canals of San Miguel, no matter the source, also belonged to the ejido. However, despite the argument in 1936 that collective ejidos were necessary because water distribution could not be handled as efficiently on an individual basis, San Miguel in 1944 had put control of irrigation water at the ultimate point of application in the hands of individual ejidatarios. It is important to distinguish the point at which each group or individual in the region had certain specified rights and duties concerning the use of water. The government had ultimate legal control of all the river and subsurface water in the region but, as noted, exercised that control only in regard to the river water. The ejido controlled the water collectively once it had entered the main canals of San Miguel and the ejidatarios decided by vote how much should be allotted to each crop. Each ejidatario then exercised individual control of the water as it entered his cotton plot, and he alone decided how much and in what manner it should be applied. Collective control of water at the point of application had proved to be not only unnecessary but detrimental to efficient production, since the collective worker was not as highly motivated as the owner of an individual plot.

The dams and canals of the Laguna region were crucial to the development of intensive agriculture in the desert climate. The property rights relating to them included the right (or duty) to build them, the right (or duty) to maintain them, and the right to control the flow of water through them. The most important part of the irrigation system was the government-owned El Palmito Dam, located in the Sierra Madre Mountains 100 miles west of Torreón. It was completed in 1944, but since its reservoir had never been full, its ability to regulate the flow of the Nazas River had not been adequately tested. Overall control of the water was vested in the Government Water District (Distrito de Riego) and the regional water boards. The primary dams and canals in the area were maintained by the government, the secondary and smaller canals by the ejidos and the private owners. Apparently the canal system had deteriorated since 1936 for lack of care, perhaps because some ejidos took the position that the government was responsible for the maintenance of all but the smallest canals and would not work on their own primary canals unless paid to do so.

Soil. In 1936, when all but 150 hectares of each large estate was either expropriated for ejidos or sold in small lots to individual owners, 1,000 of the 1,300 hectares of the hacienda of San Miguel became the property of the ejido of San Miguel, with the remaining 300 hectares going to the neighboring ejidos of Olivo and Potrero de Llano. At first all ejido crops were cultivated collectively. But the desire for the individual cultivation of crops soon became evident, and in 1941 the collective system was modified. The land was still collectively owned, but now each ejidatario was given the right to use a specified acre or two for the cultivation of his own corn crop. This modification was broadened in 1943 to permit part of the cotton crop to be grown in individual plots as well. In 1944, the ejidatarios decided that this parcela experiment should become permanent and should be extended to include the whole area of San Miguel. An engineer was hired to divide the land into separate holdings, and lots were drawn to determine the ownership of each.

The parcela owner had limited but genuine rights in the use of his land. He was not allowed to sell or rent his land if it was planted in cotton, but he could "sell" a half-grown crop of corn, in effect renting his land for a season. Further, he had to plant his parcela to whatever crop the ejido assembly specified. Almost all planting was done collectively under the direction of the work chief, though a few ejidatarios were allowed to plant the cotton in their parcelas on an experimental basis. If an ejidatario's parcela was not located in an area devoted to cotton, he had to "borrow" cotton land from a friend. Conversely, he was under obligation to "lend" part of his land if his parcela yielded more than his share of the cotton crop. This system of borrowing and lending land, which had been practiced only since 1950, was carried out with corn as well. How successful it would prove was problematical, but something of the kind was needed to prevent some ejidatarios from having only cotton and others only corn in a given year. Each ejidatario was also under obligation to work his own parcela or to find someone in his immediate family to do the work for him; old or sick men had to get special permission to hire a replacement (*remplas*). Finally, no ejidatario could leave the community and his parcela for a season without the explicit permission of the ejido.

The primary change effected by this subdivision of the land was that the incomes of the ejidatarios, all of whom now cultivated their own cotton, became proportional to the production of their parcelas. Though to some extent crop yields were subject to factors beyond a man's control, the ejidatarios felt that, by and large, yields reflected the effort and skill of the individual worker—that a lazy or unskillful ejidatario could lose money in the same year a good worker was reaping a profit.

In sum, though ownership of the soil in San Miguel was still collective in many respects and no ejidatario could buy or sell land, the partial individualization of land ownership rewarded individual effort and ability in the irrigation and cultivation of cotton. The selling of individual corn crops before harvest was not quantitatively important in the cash economy; it did, however, show that new possibilities had been created by the division of the land into individual parcelas. This pattern, if extended, could logically lead to a completely individual system of land tenure, with the buying and selling of land by individuals. If this should happen, however, the ejidatarios who lose out would almost certainly obtain the intervention of the government to recover their interest in the ejido.

Other Property. When the ejidos were created, only the soil and the irrigation canals were expropriated; all tools, machinery, and other items needed for large-scale agricultural operations had to be purchased. San Miguel had bought three tractors to do much of the work formerly done by mules, but for the most part they brought greater underemployment rather than higher profits, for the conditions that might have made them valuable were not met in the Laguna region: there were no jobs that could not be done without tractors; the labor supply was far too great for the available work; and the wages of the ejidatarios were not an expense to be deducted from the profits but a loan advanced against them, so that there was no economic incentive to decrease labor costs. The fact is, the ejidatarios of San Miguel, encouraged by officials of the Ejido Bank and impressed with the prestige attached to the ownership of tractors, overmechanized, as did many other ejidos.

About 1943, before the full individualization of the soil, the ejido decided to permit individual ownership of harnesses, plows, and small farm tools, in part because of the continued disappear-

Economic Organization 59

ance of such items from the communal storehouse and in part because of their rapid deterioration through carelessness. The ejidatarios drew lots for the community's farm implements, with the understanding that thereafter each worker would provide his own work tools. But dividing up the inventory of farm tools left most of the ejidatarios without all the tools and equipment they needed, and as a consequence a system of exchange between relatives and friends developed. If an ejidatario could not borrow a tool, he could ask the ejido to buy it for him, charging the cost against his profits at the end of the season. Tractors, harvesters, and other expensive heavy machinery continued to be community property and were cared for by the ejido mechanic. This equipment came to be used more and more on an individual basis, however. The cotton seeder, for example, was in some demand by the few ejidatarios who planted their own parcelas. Ejido trucks were almost always accessible to an ejidatario for hauling wood or for other necessary work, requiring only the *socio delegado*'s permission.

The pre-1936 dwellings of San Miguel (which had been considered of so little value by the hacienda owner that he had not asked an indemnity for them) were "owned" by their occupants who, though they had no written deed, had the right to dispose of or even destroy "their" house; they could not, however, transfer the right to the land on which the house stood. The houses built after 1936 were also individually owned, and the ejidatarios had deeds either for the dwelling or for the 20-by-20-meter lot granted to each ejidatario. A house and lot could be sold or traded to another ejidatario (with ejido permission), but the opportunities for such transactions were few. Most libres could not afford an ejidatario's house, though some had built dwellings on the house lot of a friend or relative, and a few had been given permission to build on vacant land. (This did not constitute a permanent grant: in such cases the libre had to promise not to sell or destroy his house without the permission of the ejido.) Other buildings that were not collectively owned were the small stores, the movie theater, the two pool halls, and the church.

The most valuable animals in San Miguel were the collectively owned mules, which were used primarily in the individual parcelas. An ejidatario simply asked the work chief in advance to reserve a mule for him. Little collective work was done with mules; the ejido tractors were used for most of the heavy jobs. The ejido assembly discussed the possibility of individualizing ownership of the mules in 1953, but since no practical means was suggested for dividing 82 mules among 132 ejidatarios, their care and feeding continued to be a collective responsibility.

Privately owned animals were common in the community. Almost every ejidatario had a horse or burro for transportation, as well as pigs, chickens, and goats. Some also had turkeys, ducks, rabbits, even cows. For the most part these animals were for family use, though some ejidatarios sold eggs, chickens, and goats to the cooperative or to the meat shop in San Miguel, or to buyers in Torreón.

The people of San Miguel purchased most of their food and clothing in the stores of Matamoros and Torreón, resorting to the cooperative only when they needed credit. Each family usually consumed the beans and corn the ejidatario raised on his private acreage, and in lean years, when the variable weekly wage was the family's only income, there was often no other food for months. The families that shared a household often ate together and jointly purchased and prepared the food. Some clothing was produced within the ejido. There was a tailor who specialized in men's clothing, and nearly all of the women owned or had access to a sewing machine with which to make clothes for themselves and their children. But most of the clothing, by my estimate at least 75 per cent and perhaps as much as 90 per cent, was bought in Torreón or Matamoros.

The Cotton Gin. San Miguel's cotton gin, the largest and most expensive machinery in the community, had significantly increased the ejido's income. Purchased in 1948 at a cost of 300,000 pesos (U.S. $50,000), it was the only ejido-owned gin in the region bought with ejido profits. The few other ejidos that had cotton gins had purchased them from hacienda owners; most were old and in poor condition.

Until this purchase San Miguel had sent its cotton to be ginned at Santo Tomás, as it had done in pre-ejido days when the haciendas of San Miguel and Santo Tomás were administered by the same land company. But the ejidatarios were dissatisfied with the quality of fiber produced by the antiquated machinery, and in 1945, when enough money had accumulated in their Social Fund, they began to negotiate for the purchase of a new gin. The process was a difficult one. The Ejido Bank was reluctant to release the money in the fund, arguing that there were enough gins in the vicinity (including one owned by the government), and that San Miguel's cotton production was not sufficient to amortize the "loan" in 15 years—this despite the fact that the Social Fund belonged to the ejido and no loan was involved. In the end the ejidatarios prevailed, and in the first four years of operation, 90,000 bales were ginned, far more than the 15,000 bales per year that the bank had estimated were needed to amortize the "loan." San Miguel's cotton accounted for 45 per cent of the 90,000 bales; the remaining 55 per cent was processed for 13 nearby ejidos.

The gin began as a completely collective enterprise in ownership, operation, and distribution of profits: the manager was elected by the ejidatarios; all work was collective and paid for by the day; and profits were distributed equally among all the ejidatarios, regardless of the amount of work they had done. But by 1953 libres made up over half of the gin's labor force, so that what was on the surface a purely cooperative industry, in which the workers were the owners, turned out to be on closer inspection an industry owned by the ejidatarios with a libre labor force. The libres felt themselves fortunate to obtain work in the gin, but this did not change the objective fact that in the short space of 17 years, a collective system of property designed to end the exploitation of propertyless workers had evolved into a system in which the ejidatarios profited from the labor of their propertyless libre relatives and neighbors.

Labor

Though much of the work in San Miguel was performed individually after the subdivision of the land, many jobs remained collective. This collective work was paid for by the day or by the task out of ejido funds, and was performed or directed by men selected for the job by the ejido assembly. The administrative tasks of the elected officials, described earlier, fit into this category, as did the field labor directed by the work chief. It included also certain specialized tasks performed by full-time workers, who were appointed to their jobs by the ejido assembly for varying and sometimes indefinite periods of time. In this group were the truck drivers, the wellkeepers, the mechanics, the employees of the cooperative store, the muleherders, and the caretaker of the school. The majority of these jobs were held by libres, since most ejidatarios devoted their time to their parcelas and had little inclination to accept another full-time job. One of the muleherders and three of the six wellkeepers were ejidatarios; all were older men who, by holding these relatively low-paying collective jobs, were able to increase the total income of their households and at the same time provide employment for their married libre sons in their parcelas.

The field tasks that remained collective (hence under the direction of the work chief) after the institution of the parcela system were: the preparation of the soil for all crops; the planting of cotton, wheat, and alfalfa; and the cultivation, irrigation, and harvesting of wheat (except for about 15 hectares in the 1952–53 season) and alfalfa. Ejidatarios were given preference in the collective work, but they were often busy or not interested in these jobs, so much of the collective field work was done by libres. Certain libres seemed to be given preference, though it was difficult to determine on what basis. Personal friendship with the work chief was probably a factor, since he decided which libres would work. The standard wage for most collective work in 1953 was three pesos a day; the work on private farms paid five. The ejido's work day, however, was only five or six hours, whereas the private farms

An ejidatario reinforces the furrow ridges before irrigating his cotton parcela

Cultivating a cotton parcela, a major occupation in San Miguel

Interior of the cotton gin

A central street. The house is typical of those built after 1936. Note television antennas.

One of the ejido's six wells. The structure at the right is the wellkeeper's house.

Ejido officials, June 1970

The Salón de Actos, where the ejido assembly met in 1953. The bell tower has been added since, and the hall is now used as a religious school.

The San Miguel elementary school, as seen from the Torreón-Matamoros highway

One of the school's six classrooms. The map at the back is of Coahuila.

The traditional costumes worn by the ejido girls for the fiesta of the Virgin of Refugio

Fiesta clowns and dancers

The pantomimes of the clowns are always crowd-pleasers at the San Miguel fiestas. Note the political advertisement worn by the man in the center: "Vote for Echeverría and the PRI on July 5 [1970]."

The Nicolás Robles family, with whom the author lived during his stays in San Miguel. Left to right: Nicolás, his wife, Catalina Morales Díaz, and their four daughters, María Candelaria, María Isabel, Emilia, and Josefina.

required a full eight hours of labor. Not all collective field labor was paid three pesos; irrigation work paid less, tractor driving more. In any case, the pay was adjusted to the amount of work accomplished and the degree of skill required.

Noncollective work was performed or directed by the individual ejidatario, and the wages paid for it were loans, which the ejidatario repaid to the Ejido Bank from the sale of the cotton from his parcela. Parcela labor was done by several different categories of workers: ejidatarios working in their own parcelas; substitutes (remplases) for ejidatarios or their heirs; libres who lived in the house of the parcela owner and who were not paid in cash; libres who were paid the loaned wages the parcela owner received, an illegal practice but one that was common among close relatives living in separate households; and ejidatarios who exchanged labor, working together in each other's parcelas, a form of work that was observed only in corn parcelas.

In 1953, 71 per cent of the ejido's wages (excluding administrators' salaries) was paid for individual work in parcelas, and only 29 per cent for collective work (see Table 6). In cotton, 77 per cent of the wages went for individual work in parcelas; in wheat, 79 per cent of the work was collective.

During the working season the Ejido Bank advanced money to the ejido according to the amount of work completed. The bank fixed the wage to be paid per hectare for each operation and loaned the ejido enough money each week to pay for the work

TABLE 6
Money Lent by the Ejido Bank to San Miguel for Wages and Other Expenses, 1953
(*Pesos*)

Purpose	Per hectare		Per ejidatario		
	Cotton	Wheat	Cotton	Wheat	Total
Non-labor expenses	1,477	515	2,954	645	3,599
Administrative	150	20	300	25	325
Individual work	486	24	972	30	1,002
Collective work	149	91	298	114	412
Total	2,262	650	4,324	812	5,046

SOURCE: Ejido Bank, Torreón, 1953.

that was completed. Since this loan was made on the basis of the area of land cultivated, it was relatively simple to divide the money among the parcela owners. Had the bank based the loan on mandays worked, dividing the money among the ejidatarios would have been difficult, for some worked more in their parcelas than others.

Of the 131 ejidatario parcelas in the ejido, 52 were worked by an ejidatario alone, seven by a remplas for an ejidatario's female heir, ten by a remplas for old or sick ejidatarios, and 62 by an ejidatario with the help of a libre, usually a son who lived in his household and received no cash payment for the work. The ejidatarios who paid their libre helpers in cash usually gave them the standard weekly wage; a few paid their helpers by the day, and a few also gave them a share of the parcela profits. The ejido collectively worked two parcelas, one for the school and another for the ejido mechanic.

Most of the ejidatarios did little work outside of their parcelas. The libres, however, without parcelas and without the right to work in the collective tasks, had to find employment where they could. The fortunate ones worked more or less regularly in the parcelas, the less fortunate only occasionally in collective work, or on adjoining private estates, or at nonagricultural jobs.

Labor in Cotton. In 1953 almost half of Mexico's cotton came from the Laguna region, where it has been the major crop since the 1850's. Cotton accounts for more than three-fourths of the total cultivated area and agricultural income in the region. In San Miguel it was the most important crop in terms of acreage, capital investment, and labor. The trend toward individual work, and toward the increasing use of libre labor in the cotton parcelas, was therefore of fundamental importance to the whole ejido economy.

Until 1941 all work in the ejido connected with cotton production was collective. From the initial burning of the fields to the picking of the ripe bolls, every operation was directed by the work chief and his assistants. Each worker was paid a daily wage as an advance against his share of the ejido income, based on the num-

Economic Organization 65

ber of days he worked during the year. But as noted, the work chief found it difficult to get a full day's work from some of the ejidatarios. He did not have the power to punish or reward the workers as the mayordomos had in the hacienda days; and the long-range prospect of a division of profits was not thought to be sufficient motive, since the increased income produced by the hard work of some was divided among all.

The lack of adequate rewards for effort and ability produced considerable dissatisfaction among the ejidatarios from the start. Nevertheless, there were strong forces favoring collective operations. One of these was the uncertain price of cotton, which fluctuated greatly within brief periods of time; accordingly, cotton is best marketed in large quantities by a single agent. Also, considerable capital is required in cotton production for seed, fertilizer, insecticides, ginning, and particularly labor, and it was only after 1940 that the Ejido Bank relaxed its rule requiring all crops it financed to be worked collectively.

In 1941 the ejidatarios decided to grow some of their cotton in individual parcelas, a decision made without the official approval of the Ejido Bank, and so without a loan. They decided after two seasons that individual production clearly worked better and in 1944 divided all the ejido land into permanent parcelas. The cultivation of cotton, after planting and before ginning, thus became the responsibility of each parcela owner.

Though the subdivision of the land constituted a major change in the system of collective land tenure, it did not indicate dissatisfaction with the ejido system as a whole. The important policy decisions on what would be planted, and how much, on when it would be planted, and with what techniques, continued to be made by vote of the ejido assembly. And the initial irrigation, the preparation of the land, and the planting of the seed (with a few exceptions) remained collective; only after planting did the individual ejidatario take over. Cotton picking was individual only to the extent that each ejidatario decided who would work in his parcela (usually members of his immediate family and other close relatives); the ejido assembly decided the wage that would be paid.

The weighing, ginning, classification, and marketing of the cotton were done collectively, and the ejido collectively paid all its debts and put money aside for anticipated expenses before dividing the remaining profits among the ejidatarios. The individualization of cotton operations continued after 1944, and other changes were made to reward efficient work. In 1950 a few ejidatarios were granted permission to plant their own cotton, presumably to test the possibility of decollectivizing this difficult and important operation too. Though complete individualization of operations involving tractors and other heavy machinery was not feasible, here too the ejido had taken steps to achieve a more adequate system of rewards by changing from a daily wage to a piecework basis. (The bookkeeping this required was relatively easy, since the Ejido Bank's anticipos were made on roughly the same basis, i.e., by the hectare.)

Another major change in the production of cotton was the increasing use of libres, especially in the individual parcelas. Until 1944 San Miguel's libre population was negligible, and since the ejidatarios had preference in the collective work, little of the labor in cotton was performed by libres. When they did work they were paid only the weekly wages, receiving no share in the profits when the crop was sold. Once the land was divided, however, four-fifths of the work in cotton was controlled by the individual ejidatarios. This considerably increased the employment opportunities for libres with close ejidatario relatives—and virtually closed the door on work within the ejido to those without such family ties.

In 1953 88 of San Miguel's 153 libres worked regularly in parcelas and 26 did so occasionally (Table 7). This meant that three-fourths of the libres, married and unmarried, were economically dependent on their ejidatario relatives. Much of the work in the cotton parcelas was still done by the ejidatarios, but at least one-fourth, and possibly as much as half, was done by libres.

The few libres who were paid in cash received only the weekly wages. A few ejidatarios reported that they also gave a share of the profits to their libre helpers, but this was not common, and occurred mainly when the libre did not live with the ejidatario.

Economic Organization

TABLE 7
Employment of San Miguel Libres, June 1953

Category of libres	Work in San Miguel				Work outside San Miguel			Total
	Parcela	Remplas	Collective and parcela	Non-ejido	Private farms	Non-agricultural	Students	
All libres Combined censuses	81	17	26	14	23	12	3	196[a]
Medical Service census	71	17	26	14	12	12	3	153
Married libres[b]	29	11	21	7	11	7	0	86
Single libres[b]	39	6	5	7	2	5	3	67

NOTE: The total number of libres in San Miguel was difficult to determine, in part because of their transience. Libres spent a considerable amount of time in other places—working, looking for work, or visiting relatives. The Ejido Medical Service Census listed the names of 153 libres. The Agrarian Committee listed 174 names, of which nine proved to be ejidatarios and 35 men who had left San Miguel temporarily or permanently. The total of 196 libres is based on a combined list containing names from both censuses; it does not include 29 libres who reportedly had left San Miguel permanently.
[a] Total includes 12 who left temporarily and eight who were unable to work.
[b] Medical Service census.

It is possible that the desire to make use of libre labor was one of the contributing factors in the decision to subdivide the land, though there is no proof that this was the case. In any event, the work of the ejidatarios was greatly decreased by employment of libres, and at a negligible cost.

Labor in Wheat. Wheat was San Miguel's other main crop, accounting for 40 per cent of the total cultivated area every year since 1936. Though it was not nearly as profitable as cotton per hectare (in 1950–51 wheat was the source of only 11 per cent of the ejido's gross income), it was grown in order to use irrigation water that would otherwise have gone to waste during the winter. Moreover, the Ejido Bank encouraged wheat production because the government wanted to reduce its dollar expenditures for imported wheat.

Almost all wheat operations remained collective after 1944 and, in a sense, became even more collectivized: profits after 1944 were divided equally among all the ejidatarios regardless of the production of the parcela, where before the number of days worked had determined the payment made. Only auxiliary irrigations, and two minor tasks preparatory to harvesting—leveling the irrigation borders and cutting some of the wheat by hand so the harvesting machine could enter the field without destroying part of the crop—were left to the parcela owner.

The factors that favored the collective cultivation of wheat were no greater than those that favored collective operations in cotton. In both, the use of heavy machinery and the advantages of collective marketing were important considerations. But in the case of wheat, there had been little positive incentive to decollectivize, first because the only labor required between planting and harvesting was supplementary irrigation, an operation that was not thought to affect crop yields, and second, because wheat was not a major source of income or a staple of the ejidatarios' diet and so a form of starvation insurance, as was corn.

In the 1952–53 season, however, it became clear to the ejidatarios that there were distinct advantages to individualizing wheat. In the previous year the ejido had not had enough water to irrigate half a hectare of corn for every ejidatario, as specified by the ejido assembly, so those who did not raise corn were promised half a hectare of wheat the next season, together with the profits therefrom. One result of this, the first attempt to grow wheat individually, was an extreme variation in yields: some parcelas yielded almost twice as much as the parcelas of collectively grown wheat and others only one-fifth as much. The ejidatarios, feeling that these variations were largely due to individual differences in effort and skill in irrigation, were almost unanimously in favor of changing the system of income distribution to allow the most skillful and diligent among them to earn a higher income. Meanwhile, a flaw in the collective system had also become apparent: the harvesting of the collective wheat had been delayed in 1952 because some ejidatarios failed to finish the pre-harvest tasks in their parcelas on time, and the harvesting machine was idle for several days. The ejidatarios, appreciating the seriousness of the delay and the chance that part of the harvest could have been destroyed by rain, were convinced that individual interest in a parcela's income would work against this kind of delay; and there would be less potential friction within the ejido.

The problem of how to use the collectively owned machinery in individual wheat parcelas was solved by charging each ejidatario five centavos per kilogram to harvest his crop. (At least two ejidatarios harvested their crops by hand, using the ancient method of winnowing to separate the chaff from the grain. They thus avoided the expense of renting the harvesting machine and, in addition, had the chaff to use for fodder. Most of the ejidatarios, however, did not consider the amount of money saved worth the effort.) During the harvest, the parcela owner was on hand to oversee the operation, though the men operating the harvesters knew the boundaries of each parcela. The only difference between harvesting the individual crop and harvesting the collective crop was a mechanical one: in the individual yields, the last sack had to be tied up when only partially filled, and each parcela's sacks had to be kept separate until they had been weighed.

San Miguel made an even greater departure from the collective system in 1953 in permitting several parcelas of wheat to be sold to another ejidatario shortly after they had been planted. This was tantamount to renting land, a practice that was expressly forbidden by the Agrarian Code and that the ejido itself prohibited in connection with cotton. This practice had been common for years in the corn parcelas, however, and was apparently permitted in the wheat parcelas only because they were a substitute for the corn parcelas of the previous year.

Only 10 per cent of the agricultural work for which wages were paid was in connection with the cultivation of wheat. Even with the individualization of some of the crop, 80 per cent of all the work in wheat was collective, and at least half of it was done by libres. For the most part, the ejidatarios worked in the wheat fields only when their cotton parcelas needed no attention or when a high-paying job, such as operating the harvester, was available. As in cotton, the libres were paid only the wages, which amounted to less than 2 per cent of the total ejido income. Moreover, even if the libres had been allowed to do all the collective work in wheat and to receive all the wheat wages, the ejidatarios would still have received over 80 per cent of the wheat income as profits.

Labor in Corn and Alfalfa. Corn and alfalfa were the least important crops grown in San Miguel in terms of area cultivated, cash value, and labor required. In one respect their production was similar: they were grown not as cash crops but for use within the community. But there all similarity in production ended, for corn was the most completely decollectivized crop in San Miguel and alfalfa the most completely collectivized. For that reason, a comparison of the two is helpful in determining the factors influencing decollectivization.

The traditional nature of corn production was evident in the fact that each ejidatario continued to get his seed corn from the previous year's crop instead of having the ejido buy improved seed as it did for other crops. Though less profitable per acre than cotton or wheat, corn, the staple of the Mexican diet, was highly valued as a form of insurance against unprofitable yields in those crops; this seems to have been an important factor in its early and almost complete decollectivization. Another factor was that the production of corn required no loan or credit, so the Ejido Bank had little say or control in this regard, even in the ejido's early years when any kind of decollectivization was officially opposed. Further, no heavy machinery was needed. Finally, and perhaps of greatest importance, each family consumed the corn it produced, so no marketing problem was involved.

In contrast, alfalfa was used only for collective purposes: as fodder for the ejido mules. Though there had been no major hindrances to the decollectivization of alfalfa—the need for Ejido Bank credit or for collective marketing, for example—there were no advantages to be gained by individual cultivation. Individual parcelas would merely have increased the work and responsibility of each ejidatario without increasing his income.

It was evident that the ejidatarios had no desire to decollectivize all the work that could be done individually. In each case, there was a specific advantage to the individual ejidatario in adopting a given method of production. In the case of corn, an individual crop gave the ejidatario a feeling of security; for alfalfa, collective operations meant less work for each man. In short, the ejidatarios did not value collective work or individual work as such, but only as one or the other resulted in specific and observable advantages.

The division of work in the cultivation of corn and alfalfa was much the same as in cotton and wheat: the libres did most of the collective work in alfalfa, the ejidatarios most of the individual work in corn. The exceptions were a few high-paying collective jobs (such as baling alfalfa) that were often done by ejidatarios, and libres working in corn parcelas for which there were no wages. As mentioned, corn was the only crop in which exchange labor was observed between ejidatarios.

Determining Factors in the Decollectivization of Labor. The importance of the decollectivization of labor in the economic history of San Miguel makes an analysis of the principal determinants of the process worthwhile, though it should be noted that no one factor can wholly explain why decollectivization did or did not take place. Clearly, the process has been different in each case.

The principal factor favoring decollectivization was the absence, under the collective system, of an adequate system of rewards for ability and effort. Many ejidatarios felt that they were not being paid enough for their efforts, and that others were being paid too much in comparison. The statistics on crop yields do not show whether this attitude was justified or not; however, cotton yields did not noticeably increase after 1944. True or not, the fact remains that many ejidatarios believed the original collective system to be unfair. The use of physical punishment and the threat of the loss of job and home, which had been used to motivate workers in the hacienda period, were not possible in the ejido; expulsion was possible only in the event of an absolute refusal to work. Informal types of social control, for instance, family pressure, were not not considered adequate to promote work efficiency.

The desire for a more equitable system of income distribution was strongest for the crops that had the highest value: cotton, with a high cash value, and corn, which was insurance against starvation. Because wheat and alfalfa had less commercial or other value, the incentive to decollectivize their production was not as great.

The trend toward decollectivization was strongest in connection with the operations in which individual effort and skill were thought to have the greatest effect on crop yields. It was only in 1953 that the ejidatarios discovered the relationship between skillful auxiliary irrigations and increased wheat yields, whereupon there was an immediate demand for a redistribution of wheat profits on an individual basis.

The development of a large libre group by 1944 may have been a factor influencing decollectivization of labor and the subdivision of the land. Plainly, the ejidatarios were in a much better position to use libre labor once they had their own plots: libres had to be paid in cash for collective work but could be used in parcela work without cash payment. This not only provided employment for libre sons but gave the ejidatarios more leisure time and the opportunity to find other work if they so desired.

In the case of cotton, the division of the soil occurred at the same time as the individualization of much of the labor. In the case of wheat and alfalfa, however, the existence of individual parcelas provided a continuing opportunity for a further individualization of labor after 1944. The individualization of the pre-harvest tasks in the production of wheat was possible because there were individual parcelas. Finally, the consumption by individual families of the corn crop was undoubtedly of importance in the early and complete decollectivization of labor in that crop.

Continued collective labor was favored where there was little desire for a system of income distribution that rewarded effort and ability. In the ejido's first years it was favored by the absence of a libre group. It was also favored before 1940 by Ejido Bank policies that made all loans conditional on collective work. As long as the ownership of the land was completely collective, individual labor was impossible. When the soil was partially decollectivized, what was actually decollectivized was the right to use the soil, a right that was only delegated and was subject to ejido strictures, especially on what crop could or could not be grown. The soil was thus still partially owned by the ejido, and the labor in the individual parcelas was still partially collective. The use of a crop for

Economic Organization 73

general purposes, as in the case of alfalfa, favored continued collectivization, just as the consumption of corn by individual families favored individual labor. Finally, the prohibitive costs of heavy machinery made collective ownership essential and favored collective labor in its use as well. As far as could be determined, collective marketing and credit had no direct influence on whether labor in the fields was collective or individual (except in the early years when the Ejido Bank insisted on collective labor).

A further point might be noted in discussing the subject of collectivization and decollectivization. One of the main arguments for the creation of collective ejidos, the difficulty of distributing water among some 30,000 individual ejidatarios, had not proved valid, for though San Miguel continued to negotiate collectively for its water supply and to collectively own its wells, it had successfully decollectivized water rights to the extent that each ejidatario was responsible for irrigating his own parcela. All that was required was close cooperation among the ejidatarios and coordination by the work chief so that each man was at his parcela at the right time.

It may be that there were still other factors favoring the decollectivization of labor in San Miguel. One thing was clear, however: essential to the whole process was the belief that individualization of a given operation or crop would produce increased economic rewards, or that continued collectivization was economically disadvantageous.

Nonagricultural Labor. There were a number of nonagricultural, noncollective specialized occupations in San Miguel, many of them made possible by the community's increased income after 1936. Those requiring capital (such as the ownership and operation of stores) were engaged in only by ejidatarios. The largest of these enterprises, a general store that was more completely stocked in some lines than the ejido cooperative store, was jointly owned by two brothers-in-law, who also owned some houses in Torreón, which were rented to workers in the American-owned Peñoles steel company. The community had two smaller stores (*tiendas*) that sold vegetables and staples, along with cigarettes, candy, and other small items; a third had failed when the owner overextended credit to his customers. There were also four small stores dealing exclusively in soft drinks, cigarettes, and candy, as well as three or four vendors of flavored chipped ice (*nieverías*). The nieverías needed virtually no capital to operate and made only the barest profit.

The owner of one of the tiendas, an enterprising young businessman who had lived for several years in the United States, also operated one of the ejido's two pool halls. In addition, he owned an old car that he drove as a taxi in Torreón. Another enterprising ejidatario owned the other pool hall and operated a meat shop (*carnicería*), which was open intermittently. Several years earlier this same man had started the construction of a movie theater, which, though never completed and without roof or seats, was used frequently. At first he had rented films, using his own 16 mm. projector. Later he merely rented the building to a man from Matamoros or to the traveling movie operators who came to the region from time to time. A fourth ejidatario worked daily in the *cantina* (saloon) he owned in Torreón; one of his unmarried sons assisted him there while a married son took care of his parcela. Still another important member of San Miguel's business community was a moneylender, who by virtue of his occupation was not the most popular man in the ejido but whose thriving business yielded a good profit each week and presumably would continue to do so as long as the ejido issued vouchers (worthless outside San Miguel) when the Ejido Bank was late with the weekly wages.

The small businesses of San Miguel can be conveniently divided into two groups: those operating from day to day with a capital investment of less than 100 pesos, and those with a working capital of several hundred to several thousand pesos. The first type, represented by the vendors of soft drinks, fruit-flavored ices, and other confections, were largely shoestring enterprises run by libres or women without other means of support, though one or two of the largest were run by ejidatarios. All the businesses requiring working capital—the general store, the taxi, the theater and pool halls,

Economic Organization

the cantina, the rental properties in Torreón—had been made possible by the high profits of 1950, when a record cotton crop coincided with exceptionally high world prices. The ejidatarios received an unprecedented amount of cash that year, some as much as 20,000 pesos (U.S. $2,500), enabling those who were so inclined to invest their money in a business or to enlarge an existing one.

Without exception, the substantial businessmen of San Miguel were sons or nephews of men who had had some kind of business experience in the hacienda period, indicating that despite the increased income of the ejidatarios, a tradition of entrepreneurship had not yet developed among the peons' sons.

The libres who were in nonagricultural, noncollective work were engaged for the most part in service occupations or businesses requiring little or no capital. Some worked full-time, others only part-time; some worked within the ejido and others outside it. Within San Miguel, the full-time workers included a baker, a tailor, a fruit peddler, and two goatherds; among those who worked outside San Miguel were two musicians, two mechanics, a shoemaker, and a truck driver. The only part-time workers—a shoemaker and a butcher, both of whom worked in San Miguel—spent as much or more time in collective labor as they did in their specialty. The men I term full-time workers did not necessarily work every day or all day in their occupations; however, they did no other kind of work. All told, the libres who were engaged in nonagricultural, noncollective work represented only 13 per cent of the libre population and only 8 per cent of the total male working population. But it is safe to predict that as population pressure increases, more and more libres will try to find such work outside San Miguel.

Income and Production

Cotton Production and Yields. The area of cotton planted in San Miguel between 1937 and 1953 varied from between 251 and 397 hectares, with a median of 300 (Table 8). This variation was smaller than that of many collective ejidos, most of which were not as fortunate as San Miguel in having an assured water supply

TABLE 8
Cotton Production and Yields in San Miguel and the
Laguna Region, 1936–1952
(Bales per hectare)

Year	San Miguel yields			Laguna region yields		
	Total hectares	Total bales	Yield	Ejidos	Private farms	Laguna region
1936–37	504	769	1.53	0.49	1.90	0.83
1937–38	366	814	2.22	1.43	2.02	1.61
1938–39	292	604	2.07	1.18	2.34	1.57
1939–40	275	491	1.78	1.07	2.14	1.41
1940–41	255	585	2.29	1.24	1.26	1.25
1941–42	397	663	1.67	1.22	1.98	1.68
1942–43	396	814	2.06	1.23	2.32	1.61
1943–44	350	795	2.27	0.88	1.69	1.34
1944–45	346	979	2.83	1.17	1.58	1.42
1945–46	273	774	2.84	—	—	1.79
1946–47	263	778	2.96	—	—	2.01
1947–48	261	1,100	4.21	—	—	2.04
1948–49	285	1,121	3.93	—	—	2.35
1949–50	396	1,697	4.28	—	—	2.74
1950–51	300	1,145	3.82	—	—	2.52
1951–52	300	930	3.10	—	—	2.53

SOURCE: Ejido Bank, Torreón, 1953; Secretaría de Recursos Hidráulicos 1951: 228; Cámara de Agricultura, Torreón.

from wells. Though the area of land planted in cotton has been relatively constant in San Miguel, the amount produced doubled during the 1940's, thanks to increased yields. In 1949–50 yields reached a peak of 4.28 bales per hectare and, coinciding with a record market price, resulted in a record income. San Miguel's median yield for the first seven years (excluding 1936–37)* was 2.07 bales per hectare. In the next three years, 1945–47, the first years of individual parcelas, the median yield was 2.84, a 38 per cent increase; in 1948–53 the median yield was 3.93, 90 per cent more than the yields of the first seven years (excluding 1936–37).

The determinants of the two dramatic increases in yields that occurred in 1945 and in 1948 can only be surmised. Decollectivization, begun in 1943, does not seem to have been the primary cause of the increase, for yields that year were no higher than the median

* In their first year, San Miguel and the other edjidos planted about twice as much land as could be adequately irrigated. Total production was not much affected, but yields and profits were drastically reduced.

of the previous five years. The planting of a smaller area, thus assuring more adequate irrigation, began in 1946, a year after the marked increase of 1945. No other factor has been suggested to account for the 1945 increase. The sharp rise in 1948 is concomitant with the first large-scale use of insecticides (usually sprayed from airplanes), and there is little doubt this was a major determinant of the increased yields that San Miguel had from then on. It is possible that new strains of cotton accounted for some of the yield increases, but no information was obtained on strains of cotton planted in successive years. Regardless of the reasons, it is clear that San Miguel's yields in its first eight years were better than the average yields of the region as a whole and, since 1944, higher even than the yields of the private farms.

Wheat Production and Yields. Wheat yields in San Miguel did not increase between 1936 and 1951 (Table 9). The median yields for the first, second, and third five-year periods were 1.5, 1.9, and 1.3 tons per hectare, respectively. As compared with the rest of the Laguna region, San Miguel's wheat yields were better than those of both the other *ejidos* and the private farms, whose median yields from 1937 to 1945 were: ejidos, 1.11 tons per hectare; private farms, 1.35; San Miguel, 1.62.

In 1953 San Miguel was growing some wheat in individual parcelas, and some ejidatarios believed that this would result in higher yields.

Prices of Cotton and Wheat. San Miguel's income depended not only on its production and yields but also on the prices the ejido got for its cotton and wheat. Since Mexico's cotton production was only 1 per cent of the world total, its role in determining world cotton prices was negligible. In 1936 the government bought the Laguna crop at a price above the world market price, but since then the price paid for ejido cotton has been decided by world market conditions (except in some instances when dishonest Ejido Bank officials reportedly paid the ejidos less money than the actual sale price of the crop). The price of wheat, too, was dependent on the world market, and here again, Mexican production had little impact.

In the ejido's first three years, cotton prices, like the ejido in-

TABLE 9
Wheat Production and Yields in San Miguel and the Laguna Region, 1937–1951
(Metric tons per hectare)

Year	San Miguel yields			Laguna region yields		
	Total hectares	Total metric tons	Yield	Ejidos	Private farms	Laguna region
1937	0	0	0.00	1.12	1.11	1.11
1938	195	256	1.31	1.11	1.50	1.27
1939	287	506	1.76	1.02	1.79	1.21
1940	287	431	1.50	1.25	1.80	1.42
1941	276	450	1.63	1.11	1.10	1.11
1942	260	381	1.47	0.77	1.00	0.82
1943	190	350	1.84	1.18	1.20	1.18
1944	—	311	—	0.91	1.25	1.03
1945	—	367	—	0.99	2.03	1.42
1946	201	323	1.61	1.05	1.42	1.18
1947	224	483	2.16	—	—	1.60
1948	334	530	1.58	—	—	1.21
1949	278	367	1.32	—	—	1.28
1950	—	—	—	—	—	1.31
1951	248	236	0.95	—	—	1.04

SOURCE: Treasurer's records, San Miguel; Secretaría de Recursos Hidráulicos 1951: 228; Cámara de Agricultura, Torreón.

come, were relatively low and stable. In 1941 the price of cotton rose sharply and continued upward until 1950, at which point it was almost seven times the median price for 1937–40. In the next two years the price leveled off, though it remained relatively high. Wheat prices followed much the same pattern, but since wheat accounted for only 15 per cent of the total cash income, fluctuating wheat prices did not significantly affect the total income trends in San Miguel.

Total Income. The income of San Miguel derives primarily from its wheat and cotton; about 80 per cent of its cash comes from cotton, about 15 per cent from wheat, and less than 5 per cent (a rough estimate) from work outside the ejido. Corn and alfalfa are consumed within the ejido; their combined cash value, if sold, would be considerably less than the value of the wheat. The total income from cotton and wheat comes to the ejidatarios in two basic ways: as weekly wages advanced to the ejido by the Ejido Bank and as profits when the crops are sold. Until 1948, 5 per cent of the gross value of the crops was saved and was used to buy the cotton gin. In addition, other expenses, such as farm equipment and an annual fee to the Ejido Medical Service, are deducted from the sale price of the crops. As a result, there is a considerable difference between the annual total cash income of the ejido and the money that is distributed to the individual ejidatarios and libres.

In San Miguel's first year as an ejido, the total income of the ejido was 78,200 pesos, but because of various deductions the ejidatarios received only the 48,000 pesos advanced for weekly wages (Tables 10 and 11). Wages in 1937 averaged 1.50 pesos per day, against about 1.25 pesos in 1935; thus each ejidatario's cash income in the ejido's first year was approximately 20 per cent more than it had been the previous year. In addition, the ejido as a whole had saved money in the Social Fund, had invested money in well equipment, and had paid for credit extended by the clothing store of Francisco Rodriquez in Matamoros.

TABLE 10
Total Income of San Miguel, Including Money Spent Collectively, 1937–1952
(*Thousands of pesos*)

Year	Cotton profits	Wheat profits	Wages	Total
1936–37	30.2	0	48.0	78.2
1937–38	16.7	12.6	47.5	76.8
1938–39	39.7	19.4	47.0	106.1
1939–40	4.6	10.9	47.0	62.5
1940–41	88.7	14.1	36.6	139.4
1941–42	185.8	18.6	60.0	264.4
1942–43	252.3	43.5	70.0	365.8
1943–44	205.2	52.6	80.0	337.8
1944–45	366.0	37.7	90.0	493.7
1945–46	440.9	154.7	100.0	695.6
1946–47	541.4	186.2	110.0	837.6
1947–48	759.0	139.0	120.0	1,018.0
1948–49	751.1	52.6	130.0	933.7
1949–50	1,427.8	114.5	140.0	1,682.3
1950–51	687.4	69.9	155.0	912.3
1951–52	379.8	55.0	170.0	604.8
1952–53	—	—	185.7	

SOURCE: Ejido Bank, Torreón, 1953.
NOTE: Figures for the seasons 1941–42 through 1951–52 have been interpolated from the data for 1940–41, 1949–50, and 1952–53.

TABLE 11

Total Income of San Miguel Distributed to Individuals in Cash, Excluding Money Spent Collectively, 1936–1959

(Thousands of pesos)

Year	Cotton profits	Wheat profits	Wages	Total
1935–36	—	—	51.0[a]	51.0
1936–37	−13.0	0	48.0	48.0
1937–38	−11.3	10.1	47.5	57.6
1938–39	13.9	15.0	47.0	75.9
1939–40	2.1	7.0	47.0	56.1
1940–41	56.1	10.2	36.6	102.9
1941–42	147.4	15.1	60.0	222.5
1942–43	195.5	38.9	70.0	304.4
1943–44	78.7	47.5	80.0	206.2
1944–45	249.7	32.1	90.0	371.8
1945–46	227.7	108.8	100.0	436.5
1946–47	329.2	103.9	110.0	543.1
1947–48	704.5	125.3	120.0	949.8
1948–49	625.1	−2.0	130.0	755.1
1949–50	1,427.8	114.5	140.0	1,682.3
1950–51	372.1	69.9	155.0	597.0
1951–52	205.0[b]	55.0	170.0	430.0
1959–60	0	0	879.0	879.0

SOURCE: Ejido Bank, Torreón; treasurer's records, San Miguel.

[a] Estimated on the basis of 142 workers at an average wage of 1.28 pesos per day, 356 pesos per year, which was the 1935 average of the Tlahualilo Land Company (Liga 1940: 270). In fact, the number of workers employed on the hacienda was probably about 100, and so the total wages would more properly be roughly 36,000 pesos. The 142 estimate is made in order to provide a comparison with the first year of the ejido, when there were 142 working ejidatarios.

[b] Based on an estimated 175,000 pesos spent collectively, which was the average of the five preceding years.

From 1937 to 1952 the total cash income of San Miguel increased twelvefold, primarily because of a doubling of cotton yields and a quadrupling of cotton prices. Over the same period inflation caused the cost of living to quadruple, so the total real income of the ejido, in terms of purchasing power, increased only threefold (Table 12).

Three distinct periods are discernible in the ejido's 17-year history: 1937 to 1940, a low plateau with a median annual income of 77,000 pesos; 1941 to 1947, a period of rapid increase in which the total cash income each year was 50 per cent greater on the average than in the previous year; and 1948 to 1951, a high plateau, with a median total income of 976,000 pesos.

TABLE 12
National Cost of Living Compared with Real Income of San Miguel, 1936–1967
(1937 pesos)

Year	Cost-of-living index, Mexico, D.F. (1937 = 100)	Total real income, including money spent collectively	Total real income distributed to individuals	Annual per capita real income
1936	85	—	42,000	129
1937	100	78,000	48,000	103
1938	114	67,000	50,000	104
1939	116	91,000	61,000	124
1940	117	53,000	49,000	98
1941	121	115,000	85,000	158
1942	140	189,000	160,000	278
1943	183	200,000	166,000	271
1944	230	147,000	91,000	140
1945	247	200,000	151,000	219
1946	308	226,000	143,000	197
1947	348	241,000	156,000	204
1948	369	276,000	257,000	320
1949	385	243,000	196,000	233
1950	411	409,000	409,000	465
1951	463	197,000	130,000	133
1952	530	114,000	81,000	75
1960	842	—	105,000	98
1967	979	—	90,000	92

SOURCE: National cost-of-living figures from *UN Statistical Yearbook*, 1954, 1958, 1961, 1968. Lake Success, N.Y. The total income figures in columns 2 and 3 are based on data in Tables 10 and 11. The population estimates for the last column are based on linear interpolations between data of census years.

The cost-of-living index in Mexico City rose rapidly during and after World War II, climbing 400 per cent between 1937 and 1950. As a result the median real income of the ejido from 1948 to 1951 was only 3.4 times the median income in the early years of the ejido, from 1937 to 1940. Moreover, the population of San Miguel doubled between 1936 and 1952, so that in 1948–52 the per capita real income was only 72 per cent higher than it had been in 1937–40. In 1952 a combination of drought and lower cotton prices reduced the per capita real income in San Miguel below the 1937–40 level, but no one knew at the time that this was to be the beginning of a period of reduced income, which shows no sign of abating after 18 years.

Income Variation Among the Ejidatarios. The total income of the ejido has never been distributed equally among the ejidatarios, either in theory or in practice, but the range and the variations have increased since parcelization of the land. In 1936–37 the median income from wages of the 145 ejidatarios was 312 pesos; the median annual wages of the 14 ejidatarios having the greatest income (the top decile) was 462, and the median wage of the bottom 14 ejidatarios was 237 (Tables 13 and 14). Though the ejido had profits of 30,200 pesos, none of this went to the ejidatarios as cash, so the wages advanced by the Ejido Bank represented the total cash income of each ejidatario. The median income of the top decile was 1.95 times that of the lowest decile in this year of maximal collective labor and distribution of profits. The differences were due, primarily, to the number of days that each man worked, though a few ejido officials got two days' pay for one day's work.

In 1943–44, when each ejidatario's income was based, for the

TABLE 13
Primary Sources of Income Variation Among San Miguel Ejidatarios, 1937, 1944, and 1950

	1937		1944		1950
Wages per ejidatario (*pesos*)	Number of ejidatarios (145)	Cotton parcela production (oo of kg.)	Number of ejidatarios (138)	Cotton profits (oo of *pesos*)	Number of ejidatarios (131)
0–24	2	13–20	9ᵃ	15–29	3
25–199	0	21–22	15	30–44	8ᵃ
200–224	4	23–25	12	45–59	15
225–249	7ᵃ	26–28	16	60–74	25
250–274	15	29–31	10	75–89	15ᵇ
275–299	24	32–34	15	90–104	18
300–324	31ᵇ	35–37	14ᵇ	105–119	9
325–349	15	38–40	12	120–134	9
350–374	18	41–43	11	135–149	13
375–399	10	44–46	8	150–164	5
400–424	5	47–49	9	165–179	2
425–449	1	50–52	2ᶜ	180–194	3ᶜ
450–474	10ᶜ	53–58	4	195–225	5
475–599	0	59–63	2	—	—
600–850	3	—	—	—	—

SOURCE: Treasurer's records, San Miguel.
ᵃ Median of lowest decile.
ᵇ Median of all ejidatarios.
ᶜ Median of top decile.

Economic Organization

TABLE 14
Median Income of the Top, Middle, and Bottom Deciles of
San Miguel Ejidatarios in Selected Years
(Pesos)

Year	Wages per ejidatario	Cotton profits	Wheat profits	Total	Ratio of top decile to lower decile median incomes
1936-37					
Top decile median	462			462	
Median	312			312	1.95
Bottom decile median	237			237	
1939-40					
Top decile median	1,725	2,654	480	4,859	
Median	1,106	1,701	308	3,115	2.42
Bottom decile median	712	1,095	198	2,005	
1943-44					
Top decile median	580	1,139	344	2,063	
Median	580	726	344	1,650	2.23
Bottom decile median	580	0	344	924	
1949-50					
Top decile median	1,060	18,250	867	20,177	
Median	1,060	8,750	867	10,677	3.27
Bottom decile median	1,060	4,250	867	6,177	

SOURCE: Treasurer's records, San Miguel.

first time, on the production of his own cotton parcela, the median parcela production of the top decile was 5,200 kilograms, and that of the bottom decile only 2,000 kilograms, a ratio of 2.6 to 1; in 1949-50 the cotton profits of the top and bottom deciles were 18,250 and 4,250 pesos, respectively, a ratio of 4.3 to 1. But each ejidatario received an equal share of the wheat profits and the same wages; when these amounts are added to the cotton profits, the ratio of the top decile to the lowest decreases, the maximum ratio of the four selected years being 3.27; the median is 2.32. Though parcelization was presumably undertaken in order to increase the income variation among the ejidatarios, it did not have such an effect. In 1943 the ratio was, in fact, slightly lower than it had been in 1941. The large ratio of 3.2 in 1950 was due, in great part, to an exceedingly large proportion of the ejido's total income being from cotton that year.

Weekly Wages (Anticipos) and Their Relation to Total Ejido Income

In the first years of the ejido each ejidatario was paid a daily wage of 1.50 pesos, plus a share of the annual cotton and wheat profits proportional to the total number of days he had worked during the year. Between 1937 and 1940, when this system was in effect, weekly wages accounted for 83 per cent of the annual cash income of San Miguel for the median year. With rising cotton and wheat prices between 1941 and 1944, profits increased, and only 32 per cent of the annual income of the ejido was received as wages in the median year. In the 1945-51 period, wages in the median year were only 19 per cent of the total income of the ejido. Thus, for each peso received as a weekly wage, the ejidatarios received five in cash when the crops were marketed.

The fact that most of the income of the ejido was received in January and July, when cotton and wheat profits, respectively, were distributed, worked an economic hardship on the ejidatarios. In the other ten months of the year, their real cash income was less in purchasing power than the monthly income they had received as peons before 1936. This was due in part to inflation. One way the monthly wages were supplemented was by making purchases on credit, both at the ejido cooperative and at stores in the nearby cities. In 1951 the clothing store of Francisco Rodriquez in Matamoros extended credit to 88 of the ejidatarios of San Miguel in amounts that averaged 1,210 pesos.

Table 15 shows the amount of wages each ejidatario received monthly in 1953; this amount was fixed in advance by the Ejido Bank and was contingent on the ejidatario's parcela being cultivated according to the bank's Plan of Operations. The median monthly income from wages was 75 pesos from November through March and 158 pesos from April through October. The average monthly income over the entire year was in the neighborhood of 118 pesos.

Many of the ejidatarios, unaccustomed to handling large amounts of cash, were inclined to spend their money on a feast-or-famine basis. For most of the year they did not have enough to purchase food and had to buy necessities on credit. Then for

TABLE 15
Total Monthly Wages for Individual and Collective Work in San Miguel, 1953
(Pesos per ejidatario)

Month	Amount	Month	Amount	Month	Amount
Jan.	87	May	111	Sept.	201
Feb.	96	June	138	Oct.	129
Mar.	66	July	101	Nov.	61
Apr.	163	Aug.	181	Dec.	90

SOURCE: Ejido Bank's Plan of Operations: loans for wages.

several weeks after the distribution of the cotton profits, they spent sizable sums on visits to relatives in other parts of Mexico or on other items such as radios and musical instruments. What proportion of the profits was so spent and what proportion was spent on housing, clothing, food, health, and education was difficult to judge accurately. In any case, the sums varied greatly from family to family. About 5 to 10 per cent of the ejidatarios seemed to spend their income with great care and to do an excellent job of budgeting their money. Most of these men were leaders and officials of the community, many of whom had lived in the city at some time or had come from urban families. Another group of ejidatarios, possibly as many as 25 per cent, apparently spent all their profits in a few weeks, with nothing of permanent value to show for their year's work. This group lived as poorly as the libres for the most part; indeed, a few libres appeared to have a higher standard of living. The majority of the ejidatarios, however, steered a middle course, spending part of their profits on clothes, better houses, and other substantial items.

The ejidatarios continually tried to persuade the Ejido Bank to increase the weekly wages, but without success. Whetten (1948: 234) has suggested that ejido income should be distributed more evenly throughout the year, but this would require the bank to take more interest in ejido consumption patterns and the ejidatarios' living standard than it has since 1940. The bank, content to act merely as a money-lending institution, made loans available only up to 70 per cent of the estimated value of the crop, holding that loans for higher wages would have meant an increased risk, especially in ejidos less successful than San Miguel.*

One important effect of low wages in relation to the total annual income was to encourage the ejidatarios to allow libres to do much of the ejido work, including in some instances most of the work in the individual parcelas. Though an ejidatario could not legally hire libres to do the work in his parcela, he was permitted to let libres "help" him. It was thus possible for an ejidatario to pay his own wages to hired libres and to then collect the profits as "rent" from his land.

Libre Income. After parcelization reduced the amount of collective work in the ejido, more and more libres began to work outside the ejido or in the parcelas of their relatives or friends, becoming in effect a class of hired laborers for the ejidatarios. The income of the libres of San Miguel came from the following sources (rank-ordered by total amount earned): individual parcela work; specialized permanent collective work (wellkeeping, muleherding, mechanic work); nonspecialized seasonal collective work (wheat harvesting, canal cleaning); and agricultural and nonagricultural work outside San Miguel.

In order to determine how much of the collective ejido work was done by the libres, the work records of the treasurer were analyzed for the 16-week period from February 8 to May 30, 1953 (Tables 16, 17, 18).

During the 16-week period, fewer than half of San Miguel's libres worked one day or more in the ejido's collective work, the median being ten days. Only 31 libres worked as much as 16 days in the entire period. The number of libres working in a given week ranged from 12 to 57, with a median of 19. The median income of the 103 libres who worked was 40–49 pesos per man for the entire 16 weeks, an average of about three pesos per week.

* There is some reason to suppose that the risk might in fact be decreased by such a policy. Some ejidos plant a larger area than they can adequately irrigate in order to increase their wages; if the wages per hectare were increased, these ejidos might plant a smaller area of cotton, thereby increasing their yields, and even their total production.

Economic Organization

TABLE 16
Collective Work of San Miguel Libres, February 8–May 30, 1953

Days worked	Number of libres (196)	Days worked	Number of libres (196)	Days worked	Number of libres (196)	Days worked	Number of libres (196)
0	93	4	6	8–11	8	24–27	4
1	7	5	9	12–15	16	28–34	5
2	5	6	6	16–19	6	35–44	6
3	8	7	7	20–23	7	45–79	3

SOURCE: Treasurer's records, San Miguel, 1953.

TABLE 17
Total Income from Collective Work of San Miguel Libres, February 8–May 30, 1953

Pesos earned	Number of libres (196)	Pesos earned	Number of libres (196)	Pesos earned	Number of libres (196)	Pesos earned	Number of libres (196)
0	93	30–39	5	70–79	3	140–169	3
1–9	7	40–49	10	80–99	5	170–199	3
10–19	19	50–59	5	100–119	5	200–299	4
20–29	13	60–69	6	120–139	3	Over 300	4

SOURCE: Treasurer's records, San Miguel, 1953.

TABLE 18
Weekly Employment of San Miguel Libres in Collective Work, February 8–May 30, 1953

Week	No. employed	Week	No. employed	Week	No. employed	Week	No. employed
Feb. 8	18	Mar. 8	20	Apr. 5	23	May 3	20
Feb. 15	16	Mar. 15	12	Apr. 12	15	May 10	17
Feb. 22	19	Mar. 22	57	Apr. 19	16	May 17	30
Mar. 1	19	Mar. 29	27	Apr. 26	12	May 26	38

SOURCE: Treasurer's records, San Miguel, 1953.

In addition to the 16-week analysis, a more detailed examination was made of the records for the week of June 21–27, 1953 (Table 19.) In this week, which had an unusually large amount of collective work in wheat, and hence a greater than average amount of work for libres, 42 libres were employed and earned an

TABLE 19
Collective Work and Income of San Miguel Libres, June 21–27, 1953

Wage	Number employed	Total man-days	Total income (*pesos*)
3 pesos a day	20	87	261
4 pesos a day	3	21	84
5 pesos a day	6	26	130
6 pesos a day	3	10	60
7 pesos a day	1	3	21
8 pesos a day	1	3	24
9 pesos a day	3	6	54
10 pesos a day	2	6	60
11 pesos a day	2	13	143
15 pesos a day	1	3	45

SOURCE: Treasurer's records, San Miguel, 1953.

average of 21 pesos for the week. Thus, in a very good week for libre employment, only some 20 per cent of the libres found any collective work in the ejido; and those who did made only a third of the amount of money needed to sustain a family of five at the 1936 standard of living for peons.*

Work and Residence of Libres. The majority of the 67 single libres in San Miguel, most of whom were under twenty years of age in 1953, worked in the parcela of the head of their household. Half of those who did not work in parcelas were the sons of ejidatarios, and all of these were in specialized, permanent occupations, or else were high school students in Torreón. Most of the libres who did not have ejidatario fathers worked in temporary, nonspecialized work (see Table 7).

Though the libres as a group were in a worse economic position than the ejidatarios, the single libres who were sons of ejidatarios and lived in their fathers' houses had the opportunity of working in parcelas. The others had to accept whatever work they could find; almost always this meant temporary and low-paying jobs.

Forty of the 86 married libres worked in the parcela of an ejida-

---

* Four studies made in 1936 indicated that the minimum daily budget for a peon family of five was between 1.66 and 2.97 pesos (Liga 1940: 359–60). This was equal in purchasing power to between 59 and 105 pesos per week in 1953.

tario (11 as a remplas), and 24 of the 40 lived in the household of the parcela owner. A much larger proportion of those who had permanent jobs lived in a separate house than did those who had been less successful economically. Even so, only 14 married libres lived in a separate house, and fewer than half of them owned the house they lived in. All 14 had permanent jobs, either in San Miguel or on a private farm.

Only one worker was needed to do the work in a parcela, and additional libre workers did not increase an ejidatario's income, though they might increase his leisure and his opportunity to find other sources of income. The purpose and effect of help in the parcela was to distribute the work among the adult males of the household, and an ejidatario's economic support of a married son or son-in-law was reciprocated in this way.

Still, the common residence of ejidatarios with their married sons indicated that the inferior economic status of libre sons and sons-in-law was generalized to noneconomic aspects of life as well. In 1953, most of the libres were younger than the ejidatarios, but there were enough younger ejidatarios and older libres to make it evident that the relative status of the two groups and the inequality of privilege in almost every aspect of community life were based on ejido membership, not age.

The solidarity between libre and ejidatario families within a household was probably as strong as the solidarity between the families in the joint households of the hacienda period, but it was a different kind of solidarity. It was a dependency relationship, which, though tempered by kinship and affection, was based on the inequality arising in membership and nonmembership in the ejido.

4
Social Life and the Family

Status

Equal in every way in the eyes of the law, the ejidatarios of San Miguel nevertheless quite clearly fell into four prestige classes. This ranking (which was not formally recognized by the ejidatarios themselves) seemed to be based primarily on personality characteristics, but there were other, more easily observed indices of status by which they could be differentiated. In the order of their importance to the ejidatarios themselves, those indices were: official positions held in the ejido; effort and skill as a farmer as reflected in crop yields; house type; leadership in community activities; and business activities.

The highest status group consisted of nine ejidatarios; all had been *socio delegado* or had held other high offices several times, all had crop yields that were among the highest in the ejido, and all lived in the superior type of house described earlier. In the second-highest status group were about 20 ejidatarios who had held some kind of ejido office, and who either lived in the better-type houses or were active as leaders in such community activities as the fiesta, the school, and the baseball team. Immediately below this group were the majority of the ejidatarios, about 70 in number. The men in this medium status group lived in the unpainted adobe houses, were considered average farmers, and had no profitable outside source of income. Many had held an elected office in the ejido but not a major one. About 25 ejidatarios had never held office; most of them lived in the old hacienda dwellings. Some were old or sick and had to have substitutes work their parcelas; the rest had small crop yields. Members of this group, sometimes spoken of disparagingly by the other ejidatarios, constituted the lowest status group.

Positions in the status hierarchy were achieved, not ascribed. A painted, tile-floored house, good parcela yields, and election to office only reflected certain personal characteristics, and were not in themselves an indication of status. Two former *socio delegados*, for example, were not in the highest status group because their performance in office had been indifferent; and a man who lived in one of the best dwellings was in the lowest status group because he drank to excess and was not a good farmer. His house had been virtually a gift from relatives.

Membership in a particular family did not bring status. Though two families had produced a number of community leaders, not all of the ejidatario members of these families were community leaders, or even in the high status groups. A brother of the highest-ranking ejidatario, for example, was on the borderline of the lowest group. Of the nine men of highest status, only two were from one of the two outstanding families.

No one of these indices was adequate to establish a man's position in the social hierarchy. The most reliable was probably that of ejido offices held. Income, though important, was not the primary criterion. The ejidatarios with the largest incomes were those who had other business interests. None of these men were in the highest status group; indeed, many were disparaged for their failure to work their own parcelas. In any event, their high income clearly did not put them in the highest status class, and they seldom held any kind of ejido office. One of them had been elected treasurer, and apparently did a better job than any previous treasurer, hiring a public accountant to balance and check his records when he left office. Nevertheless, he was severely criticized when a shortage showed up in the books of the man who followed him. Success in non-ejido activities led to jealousy and could lower a man's status in the community.

Age did not seem to increase a man's status, except in giving him more time in which to earn the respect of others. On the contrary, old men sometimes lost status when they were no longer able to work their own parcelas. The officials of the ejido, including the *socio delegado,* were relatively young men, usually between twenty-five and forty-five years old. There was no ruling group or advisory council of elders, though several of the older ex-officials served on important ad hoc commissions.

The Libres. The ever-increasing proportion of libres, already the largest socioeconomic group in San Miguel, constituted one of the most important changes in the social structure of the community, affecting every aspect of community life. It was therefore important to look carefully at the development of this group.

San Miguel's constitution provided that the ejido assembly could elect new members to replace those who left or were expelled from the community, as well as those who died without heirs. This was not obligatory, however. Between 1936 and 1952, only 20 men were elected to replace men who left. Most of the libres had virtually no chance of ever becoming an ejidatario, with the accompanying economic and political rights in the ejido. The libre population also included all the migrants who had come to San Miguel, most of whom were relatives of ejidatarios, as well as two former ejidatarios who had been expelled. So rapidly had the libre population grown, that in the 17 years between 1936 and 1953 it had increased in number from two to 196. About one-third of the libres were unmarried, but most were of an age to take a wife in two or three years, so their exclusion from economic and political rights could not be rationalized much longer on the basis of youth or lack of family responsibility.

A favored group of libres were the 17 remplases who worked the parcelas of seven women and ten older men. A remplas received an ejidatario's wages, plus half of the profits; he was usually close kin to the person for whom he worked, but a few were simply men who had been chosen for their ability or for other personal qualities. Also called remplas (but not so considered here) were four other libres who worked for family members, sharing their households in lieu of actual wages. Some of the remplases seemed to be as financially well off as some ejidatarios; a few had better houses and a higher standard of living. Still, they had no voice in the ejido assembly, and hence in the community's economic and political decisions. Further, their employment was conditional on the will of their ejidatario-employers: there were several libres in the community who had once held this coveted job only to lose it to another. In many ways the remplas's position was similar to that of a sharecropper, the chief distinction being that the division of income was not fixed by the ejidatario-employer but by the ejido assembly.

The numbers of remplases was certain to increase as more and more women inherited the rights of their husbands or fathers, and as more ejidatarios became too old to work. And it seemed likely that this would result in less favorable conditions of employment for the remplases—for example, a salary that consisted only of the weekly wages. In that event, the remplases would still be better off financially than the libres who worked outside the ejido, and the ejidatarios would undoubtedly have little difficulty in obtaining libres to work at essentially the same wage paid by the private estates. Another possibility was that the trend toward the use of remplases would continue to the point where the ejidatarios would collect land rent and do no work at all (though the ejido constitution expressly stated that each ejidatario had to do his share of work). As noted, several ejidatarios had already taken outside jobs to supplement their income, turning over their ejido tasks to sons or other relatives. (Most of these arrangements, however, were between members of the same household and were not in fact cash remplas arrangements.)

To a large extent the ejidatarios and the libres were differentiated by age. Only 20 ejidatarios were under thirty-three (all had been sixteen or close to that age when the ejido was established), whereas 154 libres, more than three-fourths of them, were in the sixteen–thirty-three age group (Table 20). After the original ejidatarios die, this age gap will almost certainly be closed.

Another characteristic differentiating libres and ejidatarios was migration, both seasonal and permanent. The ejidatarios had to remain in the community in order to keep up their work and their

TABLE 20
Age Distribution of San Miguel Ejidatarios and Libres, 1953

Age group	Male ejidatarios (124)	Libres (196)	Age group	Male ejidatarios (124)	Libres (196)
16–33	20	154	60–69	13	7
34–39	29	1	70–79	3	1
40–49	37	15	Over 79	1	4
50–59	18	14	Unknown	3	2

SOURCE: Ejido Medical Service Census, with additions based on San Miguel records.
NOTE: By libre I mean any male non-ejidatario resident of San Miguel over fifteen years of age.

membership in the ejido. Only 17 ejidatarios had voluntarily left the ejido, 15 or 16 of them before the land was individualized. The libres, in contrast, were frequently forced to leave the community to find employment. Moreover, they moved about freely; some left the community temporarily, and others came for short visits with relatives, so that there was a constant flow of libres in and out of the ejido.

The demographic differences between the libres and the ejidatarios, though important, were not as basic as the economic and social differences. Few libres were able to support themselves on their share of the collective work, and many worked in their father's parcela (or for some other relative), living with the parcela owner and working for him as a kinship or household obligation. Under the circumstances, the relationship was one of dependency verging on charity, for there was not enough work in a parcela to provide full employment for one worker, let alone two or more.

However small the libres' share of the ejido income, that was not the full measure of their low economic status. The libres had no right to work in the ejido, either in the collective tasks or in the individual plots; the right to work was a privilege granted them by the ejidatarios. Libres who had lived in San Miguel all their lives and who were the sons of the first settlers were as completely deprived of rights as the newest libre immigrant; at the same time, men who had only recently come to San Miguel as heirs of ejidatarios were full citizens, with all the economic and political rights attendant on ejido membership. Ejidatarios not only owned the individual parcelas of cotton and corn but were the sole recipients of the profits of the collectively cultivated wheat and alfalfa. Beyond that, the libres were expected to work when needed, even if, as sometimes happened during exceptionally busy periods, the nearby private farms were paying higher wages. (This patently unfair practice was rationalized by one ejidatario thus: "They must work when the ejido needs them because we let them live here, on *our* land, drinking *our* water.") From the viewpoint of the ejidatarios, the ejido of San Miguel had in fact lost population over the years while a number of outsiders had come there to live who had no legal right to do so, and who would be tolerated only as long as the ejido so decreed. That most of these outsiders were kinsmen, and that many had lived in San Miguel all their lives was beside the point.

In the long run, the libres' lack of political and social rights was certain to prove an even greater handicap than their lack of economic rights. Though legally the ejido was only an economic organization, the ejido assembly was in fact the civil government of the community. It sponsored and financed community fiestas, appointed the juez, and made all important community decisions. It funded the school and supervised the school budget, elected the officers of the consumer cooperative, made credit arrangements for the community with stores and doctors in the city, arranged for the community water and electricity. As the collective owner of all the land, it granted permission for the construction of houses and other buildings. In short, all social activities that required any money were under the jurisdiction of the ejido assembly where the libres had no voice. As individuals they had some informal influence on the ejidatarios, especially their relatives, but at best their influence was slight. Even in the informal loafing groups that formed every afternoon, the libres and the ejidatarios tended to sit apart, so that the informal decisions reached at such times were made without libre participation.

The libres were not allowed to use the Ejido Medical Service (even if they could afford the annual fee); nor were they granted membership and credit in the consumers' cooperative. The only community organization in which they could participate was the Society of Parents of Schoolchildren, an organization roughly equivalent to a Parent-Teachers' Association. Even then they had little say. Since the ejido allocated the land and labor for the school's parcela of cotton, and contributed additional money both to the school budget and to other school activities, the ejidatarios felt they should decide how the money should be spent. All of the officials of the society were ejidatarios. Libres seldom attended the meetings, and even more seldom spoke out. In part this was because they were younger and had fewer children of school age than the ejidatarios, but in larger part it was because of the general attitude of the ejidatario parents. That attitude was evident at a meeting of the association I attended. During a discussion of the upcoming Mother's Day fiesta, the question of whether the wives of libres were to be eligible for a raffle was debated at length. Despite the schoolteacher's indignation at the very notion of discrimination in a school-sponsored activity, it was clear that the ejidatarios considered the libre wives to be without rights in the matter; their inclusion was to be a charitable concession.

In June 1953 the libres were organized for the first time when an Agrarian Committee was formed for the express purpose of petitioning the government for a grant of ejido land. The first step in this organizational effort was made in response to a directive President Ruiz Cortínes sent to the two ejido unions in the Laguna region, ordering them to organize the libres of the entire region. The man appointed to organize the libres of San Miguel was an ejidatario who had been active as a union organizer in Torreón even before the Cárdenas regime. After approaching the ejido assembly and receiving permission to proceed, the organizer called a meeting of all the libres to explain his mission and to ask for nominations for committee officials. Unprepared for such action, the libres simply did as they were told, without discussing the purpose of the proposed Agrarian Committee, the qualifications of the leaders, or the role of the organizer, who directed and controlled all later meetings. This passivity, even indifference, was

Social Life and the Family 97

generally characteristic of the libres throughout the region; most of them reasoned that if the government wanted to make them ejidatarios it would do so in its own good time, and that their own unaided efforts would be futile.

The relationship of the Agrarian Committee to the ejido clearly reflected the low status of the libres in the community. The libres could not form an organization without the permission of the ejidatarios. Furthermore, the committee organizer was not a libre, so that an ejidatario of relatively low status dictated the committee's policies and activities. When the libres needed the ejido truck to attend a committee meeting in another community, they had to make a formal request to the *socio delegado* and, upon their return, to formally express their deep appreciation. Such formality was far from customary among members of the same family or even close friends in San Miguel, and reflected the low, outgroup status of the libres. Though most of the libres were sons or brothers of ejidatarios, there appeared to be no sense of kinship obligation between them as groups, and any help the ejido gave the Agrarian Committee was considered an act of charity rather than the fulfillment of a duty.

Apart from the obvious distinctions and status differences between libres and ejidatarios in community affairs, there were more subtle differences within the families and households. Most of the libres, whether married or single, were dependent on ejidatario kinsmen for food and shelter; most received no cash wages but worked in the parcela of an ejidatario in return for the privilege of living in his household. Few libres could afford to build a separate house; moreover, in doing so, they would have simply cut themselves off from their surest source of livelihood. The economic plight of the libre without ejidatario kin or unable to obtain parcela work was worse yet. Aside from a fortunate few who managed to work steadily at routine collective jobs that for one reason or another no ejidatario wanted, these libres had to go outside the ejido to find employment.

The libres' economic dependence on their relatives had generalized to social relationships of authority-dependency, totally unlike the family relationships in the hacienda period, when father and son or brother and brother lived together as equals, each contributing to the household income. In those days, as soon as a young peon was able to do a man's work for a man's pay, he was recognized as the social and economic equal of his father and other adults. A young libre could never achieve the same equality; his low status was ascribed, and he could not hope to attain ejidatario status, no matter how able he might be. After marriage, a young libre had little choice but to live with his father or with his father-in-law, depending on which would allow him to work in a plot.

Viewed coldly and objectively, the libres of San Miguel were simply a surplus labor group, whose help was not essential to the ejido. From the point of view of the ejidatarios, however, the libres were a very useful group. They did the low-paying collective jobs, allowing the ejidatarios to spend more time working in their parcelas; and if the ejidatarios wanted more time for leisure or business affairs, libres were always available to do parcela work too.

In the Laguna region as a whole, the libres represented an indispensable labor force for the private landowners. In the first few years after 1936, finding agricultural workers was a major problem for the private estates, which had to pay higher wages than the ejidos in order to attract ejidatarios as part-time workers. After the growth of the libre population, however, farm labor was again abundant and cheap. The situation was, in fact, even better in some ways from the viewpoint of the landowners, for the workers were now day laborers who lived in the ejidos and did not have to be cared for in slack seasons or drought years. Moreover, the new labor arrangement made it possible for the private landowners to further rationalize their operations, since they no longer had to provide such necessities as houses, schools, and water, or to offer such perquisites as subsistence plots and grazing land.

There was a growing realization among the libres that the private estates needed their labor, and some efforts had been made to organize farm labor unions. However, since the libres who worked on any given estate were usually drawn from four or more ejidos, only a union that included the libres of many ejidos could hope to conduct a successful strike.

The libres of the Laguna region also represented a source of cheap labor for industries and business in Torreón. In 1953 more than half the male working force in the Laguna ejidos was unemployed and the remainder greatly underemployed. Consequently, though many urban workers were organized into unions, the abundant supply of labor from the farms was undoubtedly a factor in depressing the wages of city workers.

The planners of the collective ejido system estimated that the amount of land required for a minimum standard of living was four hectares, and were aware that dividing a fixed amount of land by automatically including all adult males in an ejido would cancel all the benefits of land reform. But though they were conscious of the problems that would be created by a growing population, they made no provision to meet those problems. As it turned out, San Miguel was no more overpopulated than it would have been if all adult males had become ejidatarios; it was just that the economic hardship in the community was borne primarily by the libre segment of the population.

Education

After 1936 the federal government greatly increased both the number of rural teachers and the number of rural schools in the Laguna region. But San Miguel had not been content to wait for the government to put up a new building. Shortly after the ejido was established, the ejidatarios obtained a loan and built their own school, a brightly painted, three-room, U-shaped structure of a style common to other post-1936 schools in the region, with the name Escuela Federal Rural Cuitlahuac emblazoned in large letters across the front. In 1953 there were three teachers and four elementary grades, and plans were being made to build another room, so that the federal government would send another teacher for the fifth and sixth grades.

Teachers and Curriculum. Informants in San Miguel did not agree on whether the schoolteacher in the hacienda period had been paid by the government or whether the landowner had complied with the 1917 Constitution, which required "employers engaged in agricultural, industrial, mining or any other type of enterprise, whose establishments are located more than three kilometers away from the nearest town . . . to establish and support elementary schools for the benefit of the community in which they are established, provided the number of children of elementary school age be greater than twenty" (Whetten 1948: 413). In any case, the owner of San Miguel was clearly more enlightened than most of the Laguna hacienda owners, for he had established a school even before the Revolution, whereas many landowners had not done so by 1935, 18 years after the adoption of the Constitution. Still, the hacienda school had provided the children only a minimum level of literacy, plus a smattering of arithmetic (a situation that was not significantly improved in the new school's first decade).

After 1936 the San Miguel schoolteachers were paid in part by the federal government and in part by the ejido, whose constitution required the ejidatarios to cultivate a four-hectare plot (parcela escolar) collectively for the maintenance of the school. To the extent that the ejido decided on the distribution of the income from the school parcela, it exercised some control over the school. In the selection of teachers and in the determination of curriculum, however, the ejido had no say; the government could shift teachers from one place to another, regardless of the desires of the community or the teacher.

The rural schoolteachers of Mexico were assigned three major tasks by the Ministry of Education: to promote sociocultural change, to teach basic educational skills, and to instill a sense of Mexican nationalism in their students. In the broadest sense the academic studies were supposed to lead to the development of a new society; the teacher was expected not only to carry out his classroom duties but also to participate in community affairs. The first teacher in the ejido of San Miguel apparently took his role of catalyst in social change to heart, for his name appeared frequently in the minutes of the ejido assemblies; he had even signed a petition to become an ejidatario. By 1953, however, the "social" purpose of education had come to be interpreted as the sponsoring of such activities as the Mother's Day fiesta and the celebration of various national holidays. The teachers were concerned primarily with their own lives in the city and viewed their work in the ejido simply as a job, not a mission. Though there was some dissatisfaction over the teachers' aloofness from the community, most of the parents were satisfied if their children learned to read and write on a first-grade level.

Except for arithmetic, and reading and writing, the curriculum was far removed from any possible application within the ejido. Even then the level achieved was not such as to prepare students for later dealings with government officials and urban businessmen. The other subjects taught were history, general science, and grammar. History tended to be the inculcation of nationalism, with legends about the flag and national heroes; there was little evidence of a concern for the historical development of Mexico. (In an explanation of the symbolism of the Mexican flag, for instance, a teacher presented an Aztec myth as historical fact.) Science, which could have been enormously useful if the emphasis was on subjects related to agriculture, consisted of the memorization of the scientific names of objects whose common names were known to the children before they entered school. To the extent that grammar was taught, it stressed the differences between the country dialect of San Miguel and "correct" Spanish (i.e., the urban dialect of the teacher).

Diet and hygiene were also taught, and the use of toothbrushes was becoming common among the younger generation, though what role the school played in effecting this change was not certain. At any rate, much of the advice the teachers gave on diet and hygiene was unrealistic. For example, one teacher told me that she had had no success in getting the people to eat more meat. The fact was, the people of San Miguel needed no encouragement to eat meat, they simply could not afford it.

Enrollment and Attendance. Given the school curriculum's lack of relevance to the life of the community, it was not surprising to find enrollment dropping off sharply each year (Table 21). Less than half the children who enrolled completed the first grade, and only one-fifth reached the third grade.

The tendency to drop out of school after the fourth grade was apparently not connected with the fact that the children had to travel back and forth to the city every day: the dropout rate was clearly greater in the lower grades than it was between the fourth and fifth grades. In fact, most of the children who completed the fourth grade in San Miguel went on to the fifth and sixth grades.

Only four children, three boys and a girl, were enrolled in the high school in Torreón. One of the boys was quite urbanized, both in dress and in manner, and for him at least high school was likely to prove a means of social mobility. But for most of the young people of San Miguel, education did not open the door to a better life in the city—or even teach skills that might have improved their economic condition within the community.

Community Support of Education and School Activities. What control the ejido had over school activities was exercised through the Society of Parents of Schoolchildren, which approved all expenditures of the income derived from the school parcela. In 1952 that income came to 3,148 pesos. One-fourth of this amount was distributed among the teachers to supplement their government salaries; the rest, the equivalent of about U.S. $275, was the total operating budget for the school year. The ejido had a certain amount of flexibility in providing the school funds: in calculating the profits of the parcela, it could charge various expenses either to the parcela or to the ejido as a whole; it could set aside good land or poor land for the parcela; and it could tend the crop

TABLE 21
School Enrollment of San Miguel Children, 1953

Children	Grade						High school
	1	2	3	4	5	6	
Boys	31	16	8	6	5	3	3
Girls	36	13	7	3	1	2	1
Total	67	29	15	9	6	5	4

NOTE: The first four grades were in San Miguel, the fifth and sixth in Matamoros, and the high school in Torreón.

Social Life and the Family 103

with great care or let it suffer from lack of attention. Moreover, the ejidatarios had the option of underwriting additional expenses for fiestas, educational conferences, and other events related to the school program. Thus the schoolteachers were dependent to a large degree on the attitude of the ejido, not only for the school budget but even for a part of their salaries.

In 1953 the ejido's largest voluntary contribution to school activities went to the Mother's Day fiesta. Though this event was a recent innovation and directed almost completely by the schoolteachers, it had become one of the most important fiestas of the year. In 1952 the ejido contributed 1,500 pesos to this activity, an amount equal to more than half the school budget for the year.

In 1953 the Mother's Day fiesta was held on May 9. The day's events began at five in the morning, with the children, accompanied by the teachers and an orchestra hired for the occasion, proceeding from house to house, singing *mañanitas* (serenades for birthdays and other special occasions) and presenting gifts to their mothers. This was followed, in the afternoon, by an ice-cream party and raffle for all the mothers. In the evening, at about 8:30, the children gave a show for their parents in the open-air theater at the rear of the school. The program consisted of group singing and dancing, several recitations (all in praise of mothers), and two comic skits deriding husbands. The importance of the occasion was evidenced by the fact that every child had a costume made for each dance he participated in, representing a considerable expenditure of money and labor on the part of the mothers. At the conclusion of the program, a two-hour social dance ended the festivities.

It was my impression that the school was more a focus of community solidarity and pride than an institution of practical learning. The community cooperated enthusiastically with any activity like the fiesta the teachers suggested, and was willing to contribute on a grand scale for music and ice cream; but there was no discussion of improvement of the curriculum or of teaching methods by the ejido, and there appeared to be little sense of mission among the teachers.

The libre-ejidatario distinction was strong in the organization of the school, just as in all other aspects of the social organization. The ejidatarios contributed a large part of the school's expenses, including salaries, and could use this economic lever to control the teachers' activities. They dominated the parents' association and sometimes made it plain, through the school's activities, that the libres lived in the community only on the sufferance of the ejidatarios.

To the extent that the school was a stepping-stone to high school education or economic opportunity outside the ejido, it benefited very few children; in 1953 none of the libres' children were in grades beyond the second, and only four ejidatario children were in high school.

Religion

Religious belief and ritual in San Miguel were largely matters of individual conscience and practice, with almost no community pressure for conformity. The most commonly held beliefs were based on a mixture of Catholic dogma and native elements; even then religious principles seemed to be taken seriously only by the women and two or three men. Many of the ejido leaders were indifferent to religion in general and to the Catholic Church in particular, though several were active Protestants (Evangelistas) and some were avowed atheists.

The most important organized religious activities in San Miguel were fiestas honoring the Virgin of Refugio and San Isidro. These were the only community-wide, public celebrations, though Christmas, Easter, and other important events of the Catholic Church calendar were observed by individual families. Ostensibly organized to obtain a blessing for the crops, these fiestas seemed to be held primarily for entertainment. Both fiestas had been relatively more important in the hacienda period, for then they were the only community-wide organized activities. The primary expenses of the fiestas had been borne by the hacienda owner or the administrator.

The Fiesta of the Virgin of Refugio. The men who made the

arrangements for the fiesta of the Virgin of Refugio in 1953 were not elected by the ejido and were not of high status; only one was considered a deeply religious man. Except for two leaders, the dancers were young libres, whose rewards for volunteering lay in being the center of attention and participating in the feasts provided by the ejido. There was little indication that the dancers considered their performance primarily a religious ritual or obligation.

The preparations for the fiesta and the rehearsals of the *danza*, which began a number of weeks before July 4, the day of the Virgin of Refugio, provided entertainment for the people of San Miguel and served to heighten anticipation for the events to come. The fiesta itself began on July 3. The dancers, dressed in brightly colored Indian costumes and waving colored turkey feather plumes, first assembled at the long, branch-covered pavilion built a few days earlier in the central part of the village, then proceeded from house to house, collecting pictures or statues of the Virgin of Refugio and other saints.

In the evening the dancers performed almost without interruption until midnight, when they sat down to the feast that had been prepared for them. But next morning, the main day of the fiesta, they were up and around again early, carrying the holy images to the fields, running, shouting, and setting off firecrackers all the while. Upon their return at about noon, they were met near the dance pavilion by most of the men of the community, who cheered and galloped their horses in a great circle around the procession. Worn out from their strenuous morning in the fields, the dancers rested and ate in the pavilion, then resumed the dance. The people of the ejido, dressed in their best clothes, gathered in the afternoon and evening to watch the dancing, to chat, and to eat—and to admire the giant fireworks display that marked the end of the festivities. The following morning the dancers returned the holy pictures to their owners, to be saved for next year's fiesta.

Two dances are traditionally performed for the religious fiestas. I saw only one, the Dance of the Indians. The other, the Dance of the Plumes, is similar but requires different costumes. The Dance of the Indians, which represents the Aztecs paying homage to Cortez, has as its central figure La Malinche, Cortez's Indian mistress, a role danced by a young boy, dressed in white. There are several variations of the basic pattern, but in all of them the Indians line up in two rows facing each other and perform an interweaving square dance, waving their feather plumes and shouting. Nothing was said or done during the performance I saw to indicate the dance had any religious significance, except that it was performed in front of the images of the Virgin. In fact, the center of attention was often not the dance but two clowns, who imitated the dancers and sometimes comically (and even obscenely) pantomimed a baptism or a bullfight, or some other event familiar to the crowd. The chief clown was a hired performer, who lived in the nearby ejido of La Joya.

Religious ceremonies in peasant societies often come at times of the year when there is anxiety or increased leisure, or large amounts of money or harvested crops. The fiesta of the Virgin of Refugio, however, was held at a time of year when there was little leisure or money, compared with other months, and anxiety over the crop was minimal, since success more or less depended on the supply of irrigation water, a known factor. Still, however secularized it had become, there were no other organized religious events of comparable importance in San Miguel.

Formal Religion. In spite of the general lack of interest in organized religion, several men in San Miguel, urged to the task by the priest of Matamoros, had undertaken to construct a church. The ejido assembly had contributed nothing toward the building, and construction had proceeded slowly from the contribution of individual members of the community. The first man in charge of the construction, the father-in-law of a former *socio delegado*, had no official position in the ejido but was highly respected in the community. After his death in December 1952, however, an ejidatario with little status took on the job, and he had not been able to obtain either contributions of money or volunteers to work on the building. The quality of construction was poor, even by local standards of adobe masonry, and the church seemed to be

Social Life and the Family 107

deteriorating faster than it was being built. Adobes could be bought and a mason hired only as contributions were received. Some of the young libres had been persuaded to contribute their labor as *faena*, which, as commonly defined in Mexico, is labor on community tasks required of all able-bodied men; the labor on the church was neither required by the community nor devoted to community purposes. Moreover, it is doubtful that faena was ever customary in San Miguel, for in the hacienda period the community had no civil government to require it. What the man who assumed charge of the construction chose to call faena was in fact a completely voluntary religious contribution. Though most of the community felt that the church was unnecessary, since only a small number of people attended mass and there was a church close by in Matamoros, there appeared to be no real opposition to the project.

Most of the men of San Miguel were nominally Catholic, but only one claimed to go to church regularly. A good number were vocal opponents of the Catholic Church, both as an institution and as a body of belief. This group included three brothers who were avowed atheists; an extended family that belonged to a Protestant sect; a young Protestant libre without relatives in San Miguel; a former labor organizer (who believed that Jesus Christ was a Communist); and a leader of one of the largest extended families. This last man declared he was not a believer simply because he did not understand Catholic doctrine. Of the men who openly stated their disbelief in Catholicism, two had been *socio delegados*, one had been the manager of the cotton gin, and one was then a leader of the libres' Agrarian Committee.

The community's attitude toward the organized church was made particularly clear one Sunday, when the priest of Matamoros visited San Miguel during a baseball game. A friendly, personable young man and a known baseball fan, he was virtually ignored by everyone present; his greetings to various persons elicited only monosyllabic replies. And before he left he had been made the butt of a joke. When he offered to give each of the baseball players a frozen ice stick, several men and boys borrowed baseball caps and crowded around him to receive the treat. The young priest probably knew or suspected that many in the group were not on the baseball team, but he could hardly refuse them. The entire performance produced gales of laughter from the rest of the crowd, which showed no open hostility toward the priest but no respect either.

It was frequently asserted in San Miguel that the women in the community were more religious than the men. On the surface this seemed to be true, but in fact men and women were equally indifferent to their own participation in religious activities. It was only because women were responsible for making the arrangements for baptism and first communion, sacraments often insisted on by fathers as well as mothers, that one got the impression women were quite concerned with religious matters.

Despite years of exposure to the resolutely anti-religious attitude of the Central Union leaders, whose opinions were highly respected in San Miguel, the ejidatarios seemed to have been little affected in their fundamental beliefs. Even the men who were most strongly influenced by the union continued to have their children baptized in the Church and to cling to the folk-Catholic traditions.

There was no community-wide religious organization in the ejido. There were, however, two small religious groups, the Society of the Virgin of Guadalupe and the Doctrina, a Sunday school. The purpose of the society, which had 30 members (three women, 20 young libres, and seven older ejidatarios of low or medium status), was somewhat vague. According to its president, the same man who had taken over the church construction, the group was organized to serve the Virgin of Guadalupe through good works and pilgrimages. The president's wife had been chiefly responsible for forming the society and controlled the group.

The primary purpose of the Doctrina was to prepare children for their first communion. Started a few years before by the woman who organized the Society of Guadalupe, the Sunday school had at first been held in the house of the teacher; later its organizers were given permission to use the ejido assembly hall. The Doctrina was held the year around but was poorly attended except for the two or three Sundays before the first communion ceremonies. On the day I visited, the Sunday before the children were to make their first communion in the church at Matamoros, 45 children between the ages of five and twelve (about 20 per cent of the children in this age group) were in the class. The program consisted of the singing of two hymns and a group recitation of the catechism, which every new communicant was expected to memorize. Usually the school was conducted by teachers from within the ejido, but on this Sunday a woman had been sent by the church in Matamoros to help prepare the children for the special occasion.

There were no household or family patron saints in San Miguel, though pictures of the Virgin of Guadalupe, the Virgin of Refugio, or Saint John could be found in most houses, and there were statues of these holy figures in some. Nor was there a patron saint of the community. It has been suggested that Cárdenas ought to be considered the patron saint of all the collective ejidos, and indeed, every collective ejido had his picture in its office. But with all the respect and admiration for Cárdenas in San Miguel, there was no sign of a religious reverence for him. A fiesta commemorating the granting of the ejido and honoring the former president, which was held for a few years after 1936, had been discontinued; and in the years before his death in 1970, Cárdenas was not associated with any organized activities of the ejido, religious or otherwise.

Leisure

Thanks to mechanization and the availability of libre labor, the ejidatarios of San Miguel had a fair amount of leisure time, particularly as compared with the hacienda period or even the early ejido days. At the same time, the women seemed to have benefited only slightly by the introduction of such labor-saving devices as the corn mill and the sewing machine.

For the most part the members of the community spent their leisure time in nonorganized, noncommercial activities. The two major exceptions were the fiestas and baseball, both of which were formally organized, community-wide activities. From the standpoint of time, money, and participation, the annual fiestas were the most important community leisure activity in San Miguel, organized as much for sheer enjoyment as for such ostensible purposes as honoring mothers and commemorating the founding of the ejido. The people looked forward to several large fiestas each year and seemed content to add one here and another there, as long as several were held during the year. (As noted, the fiesta of San Isidro was dropped in 1953.) Of the large fiestas, only the Fiesta of the Virgin of Refugio had been held continually in the community through both the hacienda and the ejido periods. But even here, entertainment seemed to be the main goal, and the blessing of the crop only incidental.

Baseball. The ejido had supported a baseball team for many years, providing the uniforms, a playing field, and bleachers for the spectators, and furnishing the ejido trucks for the transportation of the players and fans. Other team expenses were paid for out of admission fees charged at specially organized dances during the playing season. Baseball was by no means a new sport in San Miguel: there was a hacienda baseball team before 1936, which played the teams of adjoining haciendas; but the sport had become considerably more organized since, with a regional league made up of teams from the ejidos and from industries and businesses in Torreón. San Miguel won the league championship in 1951 and had received a trophy almost annually since 1940.

The games, which were played on Sunday afternoons, were well attended: most of the men and boys of the ejido, plus a good number of visitors from nearby communities, filled the bleachers for the at-home games, and some 30 or 40 fans usually attended the away-games, accompanying the team in the two ejido trucks. Yet there did not seem to be a great deal of identification with the team; no special elation when the team won or dejection when it lost. Most of the players were young married libres, who seemed to have no special motive for participating, such as improving their status, though it is possible that the work chief, in distributing the collective work, favored libres he had noticed for their baseball skill. This was suggested by the fact that the libre who had the excellent job of driving the tractor was one of the stars of the team, as well as a close friend of the work chief. The team was often disrupted, with players leaving to look for work outside the community or quitting for little or no apparent reason; there was small social importance attached to being on the team. The spectators from San Miguel went to the games as much to be entertained as to root for their team, and they found the errors of their own players just as amusing as those of the opposing team.

The few young ejidatarios who were on the team were differentiated from the libres only in that they were more likely to have a complete uniform. There was no discrimination against libres in baseball, as there was in most aspects of the social organization. The young man who was *socio delegado* in 1953 had been a star on the baseball team a few years earlier, and the popularity he gained on the playing field was almost certainly a factor in his election to office. Although a little too heavy and out of practice, he still put on his suit occasionally and played when an extra man was needed.

The relative unimportance of organized sports to the community as a whole was suggested by the fact that an ejido basketball team, which had been active for a time, had been abandoned even though there were still regional basketball leagues. Baseball was the favorite sport of the boys, and many older men attended the Mexican major league games in Torreón, listened to the games on the radio, and kept up with the standings of the teams. Pictures of Mexican and American baseball players hung in many homes and in some of the stores. It may be that interest in the baseball team was slight only because San Miguel was trailing in the league in 1953. It was my impression, however, that the team was not an important community symbol.

Dancing. One of the most popular leisure-time activities among the unmarried boys and girls was the dance (*baile*), which was held about once a week. A form of entertainment unknown in San Miguel during the hacienda period, the baile had become a primary courtship activity in the ejido. The frequency with which the bailes were held depended to a large extent on the amount of ready cash in the community; they were held almost nightly during the cotton picking season. Most were sponsored and paid for by individual families to mark a birthday or some other special occasion.

The bailes were not community-wide activities, either in terms of participation or in terms of sponsorship. Only in the sense that one large age group of the community was involved did the ejido take official notice of them, appointing special police to maintain order.

Very few married persons attended the bailes. Some of the older men insisted they did not like the new type of dancing and music, especially the then-popular mambo, and this may have been one reason they did not attend. Usually the dance was held in the street in front of the sponsor's house. The music began at about 8:00 P.M., the dancing an hour or so later. The young people positioned themselves around a rectangular dancing area, with the girls sitting on a long bench beside the house and the boys and men lining up on either side. Eventually some of the young men, usually those who were especially good dancers or those who had *novias* (sweethearts), selected partners, and within a few minutes the dancing area was filled with couples. About midnight, couples began to drift away, unchaperoned. The dance, however, often continued until three or four in the morning, depending on the temperature and dust conditions.

The music for the bailes was furnished by a phonograph and amplifier, which had been purchased by one of the ejidatarios in 1950. He charged six pesos an hour for its use and operated it himself. The music consisted of popular Mexican ballroom pieces— two-steps, waltzes, and mambos. For 25 centavos, the phonograph operator played special requests, most often songs dedicated by one boy to another.

Loafing Groups. Mostly the ejido men spent their leisure time simply whiling away the hours together in loafing groups that congregated in the afternoons between three and sunset. These groups revealed much about the status system of the community and clique membership. The largest, composed of ejidatarios of high and medium prestige, formed in front of the ejido office. Though middle-status ejidatarios made up the bulk of the group, the high-status men—the present and past community officials—formed its core and did most of the talking. It was here that many of the ejido's policies were formulated and thrashed out before they were brought up in the ejido assembly.

Several groups of libres and low-status ejidatarios were to be found every afternoon lounging around the nieverías and in front of the barber shop. Occasionally a middle- or high-status ejidatario who lived in an adjacent house would join one of these groups, but only for a few minutes. The lowest-status group gathered in the small store of a man who was often drunk. One of his drinking companions was a cantina guitar player, whose music could be counted on to attract a number of young boys, libres, and low-status ejidatarios to the store. The baseball bleachers and the porch of the cooperative store were sometimes used for loafing, especially by the libres, but no regular groups met there. The loafing group in which a man participated, as well as his position in it, was an index to his status position in the community. Still, there were a number of men who did not participate regularly in any group. Most were either very young or very old; the older men who did not participate were likely to be of very high or very low status.

The women of San Miguel had much less leisure time than the men and found time for gossip and small talk only by combining conversation with work. The women met and talked daily at the well or at the corn mill, and often congregated in a relative's house to sew and chat. The conversation groups of the women were largely based on chance meetings or on kinship.

Reading and Radio. The majority of ejidatarios had owned radios at one time or another, but many of these were not working and others had been sold. In many cases an ejidatario bought a radio when his cotton crop brought a large profit, then later was unable to pay for repairs or batteries. The men favored broadcasts of baseball games, the women soap operas and music. Radio not only had expanded the regional and national consciousness of the people, but had clearly influenced the buying habits of the women as well. Fab detergent and Colgate toothpaste were the items most advertised on the Torreón stations, and though Fab was more expensive than other laundry products, many women in the ejido used it, agreeing, one supposes, with the observation made by a libre's wife: "It must be good because it is advertised so much on the radio."

Most of the men and women of San Miguel could read and write, but their level of literacy was fairly low, and few read much for enjoyment. Some of the men, especially the officials and past officials, read a city newspaper regularly; a few owned books. For most of the people, reading and writing were prestige skills, not skills to be enjoyed. The most common form of literature in the ejido was the comic book, especially the love comic portraying the romantic problems of the rich and the urban dweller. These love comics were even read by some of the older, high-status ejidatarios. It was difficult to determine the exact amount of reading done by the different age, sex, and status groups, but there is no doubt that the love comics were far and away the popular choice in the ejido.

Other Leisure Activities in San Miguel. Card playing, music, and checkers were among the minor leisure-time activities in San Miguel. Many men played the guitar a little, but few were proficient, and these men tended to consider music primarily a source of income rather than a pastime to be engaged in for its own sake. There had once been a band in the community but it had been dissolved: not only was the phonograph cheaper; its music was louder and more varied. Musicians were still hired for the fiestas, possibly because the music for the religious danzas was not available on records, and also because the phonograph and amplifier were not easily transported to the fields.

Only to the extent that the ejido trucks were used for transportation to and from recreational activities was pleasure driving a leisure activity in the community. The one privately owned automobile was a business investment, too expensive for casual use. Most of the men and boys were extremely interested in automobiles, however, and many knew the make and model of every car seen in the region.

Movies and billiards were the only forms of commercial entertainment within the ejido; one ejidatario had both concessions. Movies were shown at least once a week and generally attracted an audience of 50 or 60 persons. Half the films were Mexican, the rest old Hollywood Westerns. The admission was usually one peso for adults and 50 centavos for children.

San Miguel had a pool hall even before 1936, built and operated by one of the hacienda workers. It was replaced by two new buildings after 1936, the original pool hall becoming first a meeting place for the sindicato and the ejido assembly, then part of a dwelling. Only one of the two new pool rooms was still in operation. It was a gathering place for young libres, some of whom were excellent players. Nevertheless, despite the popularity of billiards in the Laguna region (every ejido I visited had at least one pool hall) only rarely was there a game in progress in San Miguel, and the pool hall owner did not appear to be making much money from his enterprise.

Leisure Activities in the City. Not only did the men of San Miguel have more leisure time after 1936, they also had more money and easier access to the city. As a result, much more time came to be spent in the city for commercial entertainment, most notably in the cantinas, in the red-light districts, and in the movie theaters.

No alcoholic beverages could be sold legally or openly consumed in San Miguel, though there was a good deal of surreptitious drinking on special occasions, such as fiestas and birthdays. In general the ejidatarios drank more than the libres, but only because they had more money. Drinking did not usually result in the open expression of sexual or aggressive impulses; its main purpose seemed to be to make the long stretches of idleness pass faster in companionable sociability. By some reports, in ejidos where liquor was sold, fighting was fairly frequent. For myself, I saw no evidence in San Miguel of drunkenness leading to fighting. At one time the ejido had permitted the sale of beer in the community, but the amount of drunkenness had increased to the point where it interfered with work, so the permission was revoked. To be sure, the saloons of Matamoros were still close at hand; however, the necessity of leaving the community plainly discouraged the men from drinking when they had work to do.

Attitudes toward drinking and drunkenness varied considerably within the community. Some of the high-status ejidatarios seldom, if ever, drank, but a few were among the heaviest drinkers in the ejido. In general the women seemed to consider drinking a male weakness that had to be tolerated. If they voiced any objection to drinking, it was only on the grounds that it seriously affected the family budget, an argument unlikely to persuade the many libres and ejidatarios who felt there was nothing worth saving money for, and that the best possible use of cash not needed for food was to buy beer and tequila.

The red-light district of Torreón had grown appreciably since 1936, and there is little doubt the increased income of the ejidatarios had contributed to that growth. The houses of prostitution of Torreón and Matamoros were especially active during the cotton picking season and after the distribution of the annual cotton profits, and were a frequent topic of conversation among the ejidatarios and libres of San Miguel. Still, the men spent far less time and money in the red-light districts than they spent in the cantinas, drinking beer.

To sum up: the leisure patterns in San Miguel clearly showed a trend toward increasing commercialization, secularization, and individualization. There had been an absolute increase in both the amount of leisure time and the amount of money available to the ejidatarios. The most important organized, noncommercial leisure-time activities, the fiestas, were as important as in the hacienda period, if not more so, but the religious emphasis had been dropped. And, finally, the differentiation between libres and ejidatarios was evident not only in the leisure activities that cost money (which was to be expected, given the income differences), but also in the participation in the loafing groups.

Though San Miguel had operated on the basis of a cash economy for almost 50 years, the notion that "time is money" had not developed in the community. The fact was, much of the leisure time in the ejido was enforced idleness, not rest or recreation; more than anything else, lack of opportunity to turn leisure time into extra income had prevented such a notion from taking root there. Indeed, a few ejidatarios had begun to pursue other economic interests in their leisure time, and it was my impression that more would do so if they had the chance.

Family Organization

As soon as the ejidos were created, changes in the community's economic and political organization were immediate and obvious. Though the changes in family organization in San Miguel had been slower and more subtle, they were nonetheless real and substantial.

Courtship and Marriage. The median age of marriage had decreased between 1936 and 1952, from twenty-two to nineteen for boys and from seventeen to sixteen for girls. In 1936 90 per cent of the boys were married by the time they were twenty-six years old, and 90 per cent of the girls by the time they were eighteen; in 1952 most boys were married before they were twenty-five, most girls by the age of nineteen. The drop in the median age of marriage was more pronounced for boys, perhaps because the economic prospects of the young libres, though not good, were considerably better than the prospects of the young hacienda peons. In any event, the proportion of unmarried adult males in San Miguel had declined significantly.

The concept of romantic love had been fostered among the ejido's younger generation by movies, the radio, and other contacts with the urban middle classes. The most eligible young men and women were those most acculturated to city ways of dress and dancing, and those earning a living in the city.

As in the hacienda days, the choice of a marriage partner was an individual rather than a family matter, though parents were now in a better position to exercise a veto because of the economic dependency of the libres on their ejidatario fathers and fathers-in-law. There seemed to be no tendency, however, to limit mate selection on the basis of the father's economic status. Few ejidatario families had income-producing property to keep in the family, and there was no special economic advantage to be gained from an alliance with other ejidatario families, since the right to be an ejidatario could not be held by both husband and wife. If a libre could work in his ejidatario father's parcela, he could gain no economic advantage by marrying the daughter of an ejidatario. Of course, if his father was a libre too, there were very definite advantages in marrying the daughter of an ejidatario; but if the youth was a good worker, he was as welcome in the household as the son of an ejidatario. Because the differentiation between libres and ejidatarios had developed so recently, and because most libres were sons of ejidatarios, no prejudices concerning the inherent worth of one group compared with the other had been formed.

Premarital sexual experience was common, judging from the prevalence of "fatherless" children, and little shame was attached to illegitimacy, though it was discouraged as an inconvenience. Trial marriage was not recognized as such, but the loose marital bonds among young married couples amounted to much the same thing. Separation was achieved by mutual consent; no formal community sanction was required. The most common type of marriage was consensual (*union libre*), involving no civil or religious ceremony. (A census taken by a nurse of the Ejido Medical Service indicated that at least 25 per cent of the marriages were consensual; I suspect the actual figure was higher.) Some prestige attached to religious marriage ceremonies, but probably no more than accompanied sponsorship of a fiesta of equal cost. In any event, the *union libre* was no less stable than marriages sanctioned by a civil or religious ceremony; all three types were easily dissolved (in the case of religious ceremonies, without Church sanction, of course). The large number of children living with grandparents or step-parents attested to the looseness of the marriage bonds. The increased income after 1936 may have increased the number of religious marriages, but there are no data by which the proportion in the pre-ejido days can be checked.

The stability of many marriages seemed to be enhanced by the low level of expectation concerning the marriage partner's role, especially among the older couples. The primary considerations were economic, and though sexual needs were satisfied through marriage, neither the husband nor the wife was expected to provide romantic love or companionship. The men tended to look to other men for companionship, and they had the opportunity for additional sexual relationships with prostitutes in the city. Despite the younger generation's exposure to the romantic ideal, I found nothing to suggest a fundamental change in this aspect of family life, though young people clearly had been influenced in the selection of a mate.

For the four students enrolled in the high school in Torreón, all of whom were children of ejidatarios, social mobility through marriage to middle-class fellow students was possible, though not probable; for the other young people of San Miguel, social mobility through marriage was not even a possibility. And, as in the hacienda period, no San Miguel child had an opportunity to mix socially with the children of the upper classes and the wealthy who, in Mexico, do not attend public schools.

Husband-Wife Relations. The degree of affection and companionship between husbands and wives appeared to be low in San Miguel, as measured by American middle-class standards. Men and women spent little time together and displayed little open affection. There were no formal or organized leisure-time activities for husbands and wives. Though occasionally a man would dance with his wife at a baile, most men regarded dancing as either a courtship activity or a practice to be engaged in with prostitutes. In any case, the bailes were primarily for single persons.

A man expected companionship and understanding from his male friends, not from his wife. Both at the bailes and in the cantinas, men often dedicated songs to one another rather than to girls, and the cantinas and other drinking situations stimulated a kind of intimate social solidarity between men never seen between husband and wife.

The looseness of the marriage ties made it difficult for the men of San Miguel to exercise an extreme degree of authority over their wives, for an unhappy wife, especially if she was young and a good worker, could always return to her parents' household or find another husband. Some women tolerated much more mistreatment than others, but there were certain limits beyond which no woman was expected by the community to stay with her husband. The prevalence of separation was ample evidence that women were not regarded as mere chattels.

The sexual division of labor was far from equal. The old saying "A man works from sun to sun, but a woman's work is never done" was only half true in San Miguel: the men were usually finished by early afternoon. Often what appeared at first glance to be the domination of a man over his wife was nothing more than the performance of duties according to a customary sexual division of labor.

To the extent that husbands exercised authority over their wives in areas not strictly defined as masculine or feminine, the domination was as strong among libres as among ejidatarios, and did not seem to have been greatly influenced by the increased economic importance of the ejidatarios. For example, I witnessed an instance in which a libre refused his wife's request to accompany her to the baile or to watch the practice of the danza (in effect not permitting her to go at all, since married women did not go to such activities alone). The man's income was small, even by libre standards, and the woman was in a position to find another husband, being relatively young and without small children to make her especially dependent on her husband. The husband's peremptory refusal can be taken as an illustration of a rather extreme degree of authoritarianism, given his relative unimportance as economic provider; at the same time, it is possible that the wife's making such a request was in itself a radical departure from tradition.

Social Life and the Family

During the Cárdenas administration an attempt was made to raise the status of the women in the Laguna ejidos by organizing them into Women's Leagues (*Liga Feminil*). Most of the groups were of short duration; the one in San Miguel, for instance, was active for only two years. Its principal undertaking had been the purchase and operation of a corn mill. The women had also met frequently to receive instruction in sewing, infant care, and other aspects of home management. But the notion of women participating actively and collectively in community affairs was too foreign to be readily accepted, and many men (as well as some women) saw the league as an attempt to take family authority away from the husband and father. In the end the group had been dissolved. One reason was that the purchase of the corn mill had aroused the opposition of a number of ejidatarios (among them, the *socio delegado*, who was reprimanded by government officials for his attitude). Probably an even more important factor was the lack of interest on the part of most of the women. By 1953 only a few of the ejidos of the Laguna region still had a Women's League, and these functioned primarily as a political arm of the Central Union.

Control of Family Finances. Few women contributed to the family income in San Miguel. Some helped in their husbands' stores, and one woman made clothes for people in the community. Apart from that, all cash income was earned by the men. Both the ejidatarios and the libres gave most of their daily (or weekly) wages to their wives, keeping aside a small amount for cigarettes, bus fare, and other small luxuries and leisure-time activities. The absolute amount of this personal expense money was usually fairly small, since the total weekly income was seldom sufficient for adequate food. In proportion to the total weekly expenses, however, it was sometimes relatively large, and was in any case much larger than the amount the woman of the house was allowed for spending money. Women did not customarily smoke or drink, and they did not go to the city as frequently as the men. For many women, a trip to Torreón to shop for food was the most important leisure-time activity of the week.

In purchasing food for the family, the women bought only staples from the cooperative store in the village, using the stores of Matamoros or Torreón where food was cheaper whenever possible. Men seemed to have more varied and more expensive clothing than women, and most of the men bought ready-made clothes in the city, whereas many of the women and children wore homemade clothes. Husband and wife often went to the city together to shop for children's clothes, but it was not unheard of for a woman to buy a dress or some other article of clothing for herself without consulting her husband about the price.

The ejidatarios did not appear to exercise any tighter control over the family expenses than the libres. The chief distinction was in the ejidatario's control over the annual profits, a source of income not available to the libres. The wives had little to say here in how the money was to be spent. Most of them were satisfied with some new clothes for themselves and the children. Even in good years, when the profits were as high as 8,000 pesos per family, the bulk of the money was spent as the man ruled: if he wished to spend it on personal pleasures—drinking, vacations, or prostitutes in the city—he did so freely, feeling that his wife had no grounds for complaint as long as the family was provided for.

Possibly one reason the ejidatarios did not demand that their profits be distributed more equally throughout the year in the form of higher wages was a reluctance to grant increased control of their income to their wives. Many ejidatarios used the exceptionally high profits of 1950 to build houses; others started or expanded private businesses. In every case, it had been the man's decision, and any benefits to the wife were only incidental. With few exceptions, the kitchen, where the wives spent most of their time, remained the darkest and worst-ventilated spot in the house. Kerosene stoves, running water, and other conveniences had been installed in only a few homes, even though these "luxuries" were well within the financial reach of most ejidatario families.

Parent-Child Relations. In the hacienda period, children were the only form of old-age security the peon had; though many young people were forced to migrate in search of jobs, one or two children remained at home to work with their parents and to share their household. Status within the family was based on the ability to work and produce income; a boy became his father's peer as soon as he contributed a full share to the family income, usually before he was twenty-five years old. This attitude was still strong in San Miguel in the sense that the community leaders were fairly young. (The secretary in 1953 was only twenty-eight; the *socio delegado* was thirty-nine, and his predecessor was thirty-two when he took office.)

The children of the ejido were expected to obey their parents, but obedience was not considered a cardinal virtue. There was no success ideal driving parents to demand that their children learn expected roles very early or skills especially proficiently. A few ejidatarios had sent their children to high school in the city to be trained for higher social and economic positions, but none of the children had yet "succeeded." The majority of the boys learned to grow cotton and wheat by working along with their fathers, and the work thus learned was neither a task nor a duty; it was the customary role of a man.

The main change in the father-son relationship after the creation of the ejido was an objective economic change, and there was little evidence that the old attitudes had been replaced by views consistent with the new situation. Since each ejidatario's income was fixed for life (within certain limits that were independent of the number of workers in his family), children, far from representing old-age economic security, could not even increase the household's income unless they found employment outside the family's parcela. The ejidatarios were permitted to let a son (or son-in-law) do their parcela work when they were too old or sick to carry on, but from the standpoint of pure economics, a hired libre would serve the purpose equally well.

A married libre son was not a financial equal in an ejidatario's household; he was an economic dependent. In the hacienda period, a strong economic tie between parents and children was added to the bonds of affection; in the ejido period the bonds of affection were often strained under the burden of economic dependency.

Interfamily Relations and Artificial Kinship. Two-family households other than those composed of extended families were uncommon in San Miguel. As far as I could determine, only two ejidatario and four libre families shared homes with friends, and even then it is possible there was a real or quasi-kinship relationship that I did not discover.

There was little visiting between families as groups; most social interaction was among the men in the loafing groups, and among the women in the course of their work. Since the community was so small, distance played little part in determining participation in visiting groups, though because of the tendency for the various extended families to live close together, proximity reinforced kinship ties.

Even though San Miguel had existed as a community less than 50 years, the network of kinship was growing rapidly; it was the exceptional person who was not related by marriage or blood to many people in the community. This was especially true of the ejidatarios and their families, many of whom had lived in the community for most of their lives. Moreover, natural kinship ties had been extended to some degree by ritual kinship, notably by *compadrazgo,* or godparenthood. This institution, which serves to increase the security of the individual and the solidarity of the group, was reported to be important in the ejido of Cuije in the Laguna region, but it was neither strong nor extensive in San Miguel. Most of the adult males over thirty had at least one *compadre,* and some as many as three or four. The choice of godparents for a child was generally made by the wife, however, and many men could not even remember the names of their compadres, especially if they lived in another community. In any event, the term compadre was often applied loosely, in a half-joking fashion, to any close friend.

In a community composed of persons who were essentially all strangers only one or two generations before, one might reasonably expect to find the development and strengthening of ritual kinship in the absence of genuine kin ties. At the very least one would expect to find that there had been a strong need for artificial kin ties early in San Miguel's history, and that the need for such institutions as compadrazgo had lessened only as real kin ties were extended. But there was no evidence to support this view: the compadre bonds seemed as few and as weak among the older men as among those born in the ejido period. The only explanation of the weakness of the compadrazgo system in San Miguel was the corresponding weakness of organized religion in general, and the unimportance of baptism in particular. True, most children were baptized, but apparently not so much to save their souls as to uphold the social status of the family; there was no community sanction against those who did not have their children baptized.

Kinship, Residence, and Work. The number of marriages formally recorded in San Miguel between 1936 and 1953 was too small to permit hard-and-fast conclusions concerning rules of residence for newly married couples. Nevertheless, certain tendencies seemed clear, the most important of which were supported by statements of informants. The cultural ideal was for a young couple to reside with the groom's parents or in an adjoining or nearby house, but a good deal depended on the economic position of the parents. Residence was always with an ejidatario parent, regardless of whether he was the groom's father or the bride's. If the fathers of both were ejidatarios, residence was likely to be with the groom's father, unless the household was already overcrowded and the bride's parents were in a better position to take the couple in. On occasion a relative of the groom might have an extra house that the couple could use or would help the young man build a house, in which event the young man and his wife lived with the groom's parents until the new house was completed. To illustrate just how strong the tendency for patrilocal residence was, one young libre inherited a house and the right to be an ejidatario from a family friend but remained with his father, permitting a friend to use the house he had inherited. And as further evidence of the overriding importance of the economic position of the parents, some young couples shuttled back and forth, living for a time with the young man's ejidatario father, then with the wife's, moving whenever one of the ejidatario fathers gave the young husband parcela work.

Once the ejido was established, the ejidatarios (with few exceptions) moved out of the old hacienda dwellings and built new, larger, and better houses. Many fathers, sons, and brothers built adjoining houses, though they continued to share their income and to eat together. This was especially true in the first few years. Later, with the growth of the libre sector, the old pattern of joint residence began to reappear, this time however on the basis of economic dependency rather than mutual support. At least 20 per cent of the ejidatarios shared their houses with married sons, and 12 per cent with married daughters and their husbands.

Most married libres lived with their ejidatario fathers or fathers-in-law; about 20 per cent lived in separate houses owned by ejidatario relatives, and several had managed to build their own one-room, dirt-floored, adobe dwellings. The residence pattern of the remplases reflected their better economic position: 14 lived in separate houses and only three lived with the ejidatario for whom they worked.

The second type of joint residence and joint work, that of ejidatarios living together on an equal footing, was common in the ejido's first few years. But that type of egalitarian household was becoming rare. In the mid-1940's more than 50 ejidatarios worked their land jointly with an ejidatario father, son, brother, or other close relative, and only about 20 ejidatarios with close ejidatario relatives did not work their parcelas jointly. Of the remaining 70 or 80 ejidatarios who had no close relatives with whom they had built up a pattern of joint residence during the hacienda period, only two worked jointly with non-kin ejidatarios, and then for only one year.

By 1953 the number of ejidatarios who worked their parcelas jointly with another ejidatario had decreased by almost 50 per cent, indicating that this practice was a carryover from the hacienda period rather than a custom developed and encouraged by ejido conditions. From a practical standpoint, joint income is difficult to divide equitably unless all living expenses are shared, and with the construction of many new houses, joint consumption became increasingly rare among ejidatarios. Most ejidatarios, when questioned on the reason they had stopped working their parcela jointly with an ejidatario relative, replied either that they had separated when they married and began living apart, or that their partners had not done their share of the work. They left unmentioned the growing number of libre sons and sons-in-law who needed to work; accommodating a libre dependent necessarily broke up the joint parcela arrangements with ejidatario relatives.

Part II
San Miguel in Perspective

5
San Miguel Since 1953

Physical, Demographic, and Socioeconomic Changes
When I returned to San Miguel in 1966, after being away for 13 years, what struck me most forcibly at first was how little it had changed in appearance; I could almost believe that I had not been gone more than a few days. Except for a few new adobe houses, most of them built by married men on their fathers' house lots, the pattern of dwellings was virtually unchanged. Nor were there any new nonresidential buildings or major additions to old ones apart from two classrooms that had been added to the school and an adobe tower that had been built on a corner of the Salón de Actos, which had been converted into a Sunday school. There were a few more shops selling soft drinks, cigarettes, pastries, and fruit; a government-operated CONASUPO cooperative had replaced the ejido-controlled consumer cooperative; and the residences evidenced an increasing differentiation of life styles and unequal access to additional sources of income outside the ejido. The community-owned mule herd had dwindled, and horses, a prestige item in San Miguel, were fewer in number; still a new and expensive (U.S. $10,000) Caterpillar tractor had been purchased in these years. Notwithstanding my first impression, I found that San Miguel had managed only to meet its labor expenses over the last ten years, which explained the lack of new buildings. On closer inspection, there were several signs of deterioration. For example, the ball park no longer had bleachers; and whereas in 1953 San Miguel had had electricity for individual homes, now it had electricity only for the school, the cotton gin, and the well pumps.

TABLE 22

Rural and Urban Population of the Municipios of Matamoros and Torreón, 1921–1960

	Matamoros				Torreón				San Miguel	
Year	Rural population	Rate of rural increase per decade	Urban population	Rate of urban increase per decade	Rural population	Rate of rural increase per decade	Urban population	Rate of urban increase per decade	Population	Rate of increase per decade
1921	13,959		4,549		5,547		50,902		284	
1930	14,277	2%	6,001	32%	8,905	60%	66,001	30%	329	16%
1940	20,553	44	7,961	33	11,969	34	75,796	15	502	53
1950	27,470	34	10,154	28	18,262	53	128,971	70	879	75
1960	32,861	20	13,770	36	23,252	27	179,901	40	1,065	21

SOURCE: Secretaría de Economía, Censo General de Población, 1930, 1940, 1950, 1960. Mexico, D.F.

San Miguel Since 1953

Population. When I began looking for the ejidatarios and libres I had known best 14 years before, I learned that many had left San Miguel, including some of the most able and successful men in the ejido. The work chief, Florentino Reyes, had gone to Ciudad Obregón to be a mechanic; Aurelio Gonzalez, one of the high-status libres, had gone to Juárez to work; Teodoro Argumaniz, the first *socio delegado* of the ejido, was now the manager of a private farm about 20 miles away and was seldom in San Miguel. My suspicion that San Miguel's population boom of the 1930's and the 1940's had slowed down was confirmed by the 1960 government census count of 1,065 which, though 186 more than the 1950 census, indicated an increase during the 1950–60 decade of only 21 per cent, as compared with increases of 53 per cent between 1930 and 1940 and 75 per cent in the next decade.

The number of ejidatarios had been reduced to 90 by death, emigration, and expulsion, and the libres now outnumbered them by a ratio of two to one. As compared with the Ejido Medical Service's census of 1952 (supplemented by my count of libres) there had been a decrease of 22 persons between 1952 and 1960. The large net immigration into San Miguel from 1935 to 1950 had been reversed, and during the 1950's, people left San Miguel in numbers that may be estimated as being between a minimum of 9 per cent per decade (based on the 1950 and 1960 censuses) and a maximum of approximately 30 per cent (based on the rate of natural increase and the difference between the 1952 and the 1960 census).

That this exodus from San Miguel was not unique, but was representative of a process that took place throughout the region, is evidenced by the rural and urban population growth rates of the municipios of Matamoros and Torreón, shown in Table 22. The growth rate of the rural population of the municipio of Matamoros, of which San Miguel is a part, fell from an average of 39 per cent between 1930 and 1949 to 20 per cent in the 1950's. In the adjoining municipio of Torreón, the rural population increase per decade fell from 43 per cent to 27 per cent in the same time period. Simultaneously, the rate of urban population increase in the town of Matamoros rose from 28 per cent in 1940–50 to 36 per cent in 1950–60.

In Mexico as a whole the proportion of rural inhabitants in the population has been steadily decreasing during this century, falling from 82 per cent in 1910 to 66 per cent in 1930, and to 61 per cent in 1960. For two decades the collective ejidos had reversed this trend for the Laguna region; they had not only retained their residents but also attracted a sizable proportion of immigrants from other rural areas. Since the 1950's, however, San Miguel and the other collective ejidos of the Laguna region have been filled beyond their economic capacity; they are now contributing as much to the population of the already overcrowded towns and cities as are the other rural areas of Mexico.

Collective and Individual Work of the Ejidatarios. Since 1967 there has been a further decollectivization of work and differentiation of income distribution in San Miguel, but the changes have not been comparable to those that occurred when the land was divided into parcelas in 1942. The planting of cotton in the parcelas has become the responsibility of each ejidatario, and the recommendation that crops other than cotton and wheat be grown by the ejidos, as proposed by economists and agronomists since the 1930's, has finally been accepted. In 1968 San Miguel planted 20 hectares of grapes, which were cultivated collectively; they yielded, in 1969, some 60,000 kilograms of grapes, replacing wheat as the ejido's second crop. Japanese wine manufacturers have contracted for a large part of the grapes of the Laguna region, and there is widespread belief that the region has found a way to break its dependence on cotton and wheat.

One trend that has continued as far as it can is the use of libre labor in collective tasks. Libre sons, sons-in-law, and other kin are now doing almost all the collective work, which is distributed equally among the dependents of each ejidatario.

Income. The income drop San Miguel experienced in the two years following the record high of 1949–50 continued, and from 1957 until my return in 1967 there had been no income other than the wages advanced by the Ejido Bank and the small amount earned outside the ejido. The reasons for this were not readily apparent, and the officials of both San Miguel and the Ejido Bank were reluctant to discuss the situation or to provide data. There were conflicting opinions among the ejidatarios about what had caused the decline—factors beyond the control of the ejido and the bank, such as inflation, the lowered water table, and low cotton prices, or simply incompetence and graft.

Though detailed data were obtained only for 1960, sufficient evidence was gathered to make it clear that two major factors account for the decreased income of the ejido: cotton prices, which remained at the level of the early 1950's, and inflation, which doubled production costs.

Between 1954 and 1967 the prices paid for Mexico's exported cotton in pesos remained virtually unchanged; though there were small fluctuations, the prices remained remarkably stable, particularly in the 1961–67 period when the highest price was the 1967 figure of 665 pesos per metric ton and the lowest was the 1962 figure, 642 pesos per metric ton. (UN Yearbook of Trade Statistics 1969: 570.) As a result of inflation, however (see Table 23), real cotton prices decreased steadily and were only half as high in 1967 as in 1954.

TABLE 23
Wholesale and Consumer Price Indexes, Mexico City, 1936–1967

Year	Index	Year	Index	Year	Index	Year	Index
			WHOLESALE	PRICES			
1936	84	1944	178	1952	402	1960	598
1937	100	1945	199	1953	394	1961	604
1938	105	1946	229	1954	429	1962	610
1939	100	1947	242	1955	489	1963	—
1940	102	1948	260	1956	512	1964	640
1941	109	1949	288	1957	536	1965	653
1942	121	1950	311	1958	560	1966	665
1943	146	1951	386	1959	567	1967	684
			CONSUMER	PRICES			
1936	85	1944	230	1952	530	1960	842
1937	100	1945	247	1953	520	1961	851
1938	114	1946	308	1954	546	1962	859
1939	116	1947	348	1955	634	1963	—
1940	117	1948	369	1956	666	1964	876
1941	121	1949	385	1957	702	1965	911
1942	140	1950	411	1958	780	1966	945
1943	183	1951	463	1959	801	1967	979

SOURCE: *UN Statistical Yearbook*, 1954, 1958, 1961, 1968. Lake Success, N.Y.

The total cash production costs of the ejido are equal to the total amount borrowed from the Ejido Bank less the monthly wages, which are not expenses but advances against the eventual profits. In the first five years of the ejido's operations the median annual production costs (in 1937 pesos) were 84,800 pesos; in the succeeding five-year period, 1941 to 1945, they decreased to a median of 45,400 pesos (Table 24). In the five years after the purchase of the cotton gin in 1948, production costs rose to an annual median of 135,100 1937 pesos, but during this period the ejido's real income was higher than for any other five-year period in its existence. Between this prosperous period and 1960, production costs rose to 180,100 1937 pesos, an increase of 32 per cent. This means that during this period the increased production costs were only partially accounted for by inflation; there was a 32 per cent in-

TABLE 24
Total Operating Expenses of San Miguel, 1936–37 to 1959–60
(Thousands of pesos)

Year	Total bank loan	Wages	Non-labor production expenses	Non-labor expenses (1937 pesos)
1936–37	180.5	48.0	132.5	132.5
1937–38	136.6	47.5	89.1	84.8
1938–39	133.8	47.0	86.8	86.8
1939–40	120.8	47.0	73.8	72.3
1940–41	120.9	36.6	84.3	77.3
1941–42	115.0	60.0	55.0	45.4
1942–43	179.7	70.0	109.7	75.1
1943–44	145.9	80.0	65.9	37.0
1944–45	180.6	90.0	90.6	45.5
1945–46	171.0	100.0	71.0	31.0
1946–47	88.3	110.0	−21.7[a]	−9.0[a]
1947–48	376.1	120.0	256.1	98.5
1948–49	439.9	130.0	309.9	107.6
1949–50	987.5	140.0	847.5	272.5
1950–51	967.0	155.0	812.0	210.4
1951–52	713.0	170.0	543.0	135.1
1959–60	1,956.0	879.0	1,077.0	180.1

SOURCE: Treasurer's records, San Miguel, 1960. The 1942–52 wage figures have been estimated by interpolation between 1941 and 1953 data. The figures in the last column are based on the wholesale price index in Table 23.

[a] The 1946–47 data are not included in the calculation of the five-year medians mentioned in the text: either the bank loan figure or the estimate of monthly wages is incorrect, since the monthly wages obviously could not be greater than the total bank loan.

crease over and beyond what could be accounted for by the decreased purchasing power of the peso.

Caught in a squeeze between rising production costs and falling cotton prices, the ejido was given relief by the Ejido Bank in the form of larger loans for weekly wages. From 1948 to 1952 the median total income of the ejido (including money spent for collective purposes) was 923,000 pesos, of which 150,000 was in the form of wages. By 1960 wages for all purposes had been increased to 879,000, but this was the entire ejido income; there were no profits. The Ejido Bank increased its risk-taking by lending this much money for wages when production costs had more than doubled and cotton prices had not increased at all. The 1,956,000 pesos the Ejido Bank lent to San Miguel in 1960 (Table 25) was as much

TABLE 25
Uses of Ejido Bank Loans to San Miguel, January–December 1960
(Thousands of pesos)

Use	Amount	Use	Amount
Labor (paid to San Miguel workers)		Consumables	
Wages, parcelas, and collective work	502	Cotton seed	29
		Wheat seed	26[c]
Cotton picking	164	Alfalfa seed	4
Cotton gin	94[a]	Electricity	223
Preparation of land for planting	77	Insecticides	199
		Fuel and lubricants	77
Administrative salaries and miscellaneous	18	Fodder	14
		Baling wire and cloth	93[a]
TOTAL	879	TOTAL	665
Taxes, interest, etc.		Services (of outside workers)	
Taxes	75[b]	Machinery repairs	92
Interest	82	Application of insecticides	61
Insurance	71	Classification of cotton	6
Water commission	16	Construction	2
Medical service	7		
TOTAL	251	TOTAL	161
		GRAND TOTAL	1,956

SOURCE: Treasurer's records, San Miguel, 1960.

[a] The total amount for baling and ginning cotton, 187,000 pesos, was broken down into 94,000 pesos for labor and 93,000 pesos for materials on the basis of the 1948 cost percentages for materials, labor, and profits.

[b] Estimated tax = roughly 4 percent of gross income = roughly 4 per cent of bank loan in 1960.

[c] It is possible that this 26,000-peso item is for something other than wheat seed, but this seemed to me the most likely use of this sum, which was not labeled in the records.

money as San Miguel had grossed from cotton and wheat together in the record year of 1949–50; it was approximately 25 per cent more than the ejido had grossed in the four good years between 1948 and 1952 (excluding 1950), when there had been a comfortable margin between gross income and production costs.

Standard of Living. The peso income of San Miguel was not much lower between 1952 and 1966 than it was during the five best years, 1946 to 1950, but inflation steadily reduced the purchasing power of the ejido, and the per capita real income was further reduced by the population growth. The total real income of the ejido, including money spent collectively, decreased from a median of 130,000 1937 pesos in the 1946–50 period to 105,000 1937 pesos in 1960, a 20 per cent decrease (see Table 12). The population growth combined with inflation decreased the annual per capita real income to 75 1937 pesos in 1952 and 98 pesos in 1960, as compared with a median of 233 pesos in the high years of 1946–50, and 104 in the first five years of the ejido.

The meaning of San Miguel's income decrease may be better understood if it is compared with the distribution of income among other Mexican families. Data from two studies of the income distribution of Mexican families, made independent of one another in 1956 and 1957 (I. Navarette 1960; Cline 1962), are given in Tables 26 and 27. By comparing San Miguel's per-family income with these studies, one may see the extent to which the

TABLE 26
Income Distribution of 5,000 Mexican Families, 1950 and 1957, by Percentiles

Percentile	Average monthly income in 1957 pesos		Change in real income	Percentile	Average monthly income in 1957 pesos		Change in real income
	1950	1957			1950	1957	
1–10	247	192	–.20	51–60	504	632	+.25
11–20	311	304	–.02	61–70	641	835	+.30
21–30	348	350	.00	71–80	788	1,128	+.44
31–40	403	429	+.06	81–90	989	1,658	+.68
41–50	440	485	+.10	91–100	4,450	5,420	–.50

SOURCE: I. Navarette 1960: 165.

TABLE 27
Income Distribution of 5,002,100 Mexican Families, 1956

Monthly income in 1956 pesos	Per cent of families	Monthly income in 1956 pesos	Per cent of families
Below 76	0.0	600–799	16.9
76–149	4.3	800–999	8.8
150–199	17.2	1,000–1,499	9.6
200–299	17.5	1,500–2,999	1.9
300–399	11.9	Over 3,000	1.5
400–599	10.4		

SOURCE: Cline 1962: 344. Taken from *Ingresos y egresos de la población*, Mexican Government census.

TABLE 28
Per-Family Average Monthly Income of San Miguel in 1950 and 1960 Compared with Income of Other Mexican Families

	1950 San Miguel income[a] (882,000 1950 pesos/year)		1960 San Miguel income (879,000 1960 pesos/year)	
Category	Per ejidatario family (131)	Per family (212)	Per ejidatario family (125)	Per family (252)
1937 pesos	136.5	84.4	70	34.7
Equivalent 1956 pesos	909.1	561.8	466	231
Equivalent 1957 pesos	958.2	592.2	491	244
Percentile rank among Mexican families (1956 pesos)	82[b]	59[b]	54[b]	24[b]
Percentile rank among Mexican families (1957 pesos)	84[c]	62[c]	49[d]	10[d]

SOURCE: Cline 1962: 344; I. Navarette 1960: 165.
[a] Based on average of 1948–52 income.
[b] Based on 1956 income distribution of five million Mexican families (see Table 27).
[c] Based on 1950 income distribution of five thousand Mexican families (see Table 26).
[d] Based on 1957 income distribution of five thousand Mexican families (see Table 26).

ejido's income has declined in comparison with the rest of the Mexican population. Two kinds of comparison are shown in Table 28. The total ejido income divided by the number of ejidatarios gives an idea of what the economic status of San Miguel might be if the community did not have the problem of landless libres. The total ejido income divided by the total number of families, libre and ejidatario, gives an idea of the average family income level (without however indicating the extreme variation between the poorest libres and the most affluent ejidatarios).

The average annual income of San Miguel for 1950 (and for the two years just before and after) was 882,000 1950 pesos. For the 131 ejidatario families, this represented a monthly average income of 136 1937 pesos, 909 1956 pesos, and 958 1957 pesos. When all families are counted, the respective figures drop to 84, 562, and 592. For this five-year period the average San Miguel family income (including libre families) was higher than the income of 61 per cent of all Mexican families. This is an amazingly high relative economic position, especially when one considers that the ejido was supporting 81 libre families in addition to its 131 ejidatario families, or 41 per cent more families than the 150 provided for in the ejido grant. If the 81 libre families are not included in the calculation of the average family income, the San Miguel average family income would have been at the eighty-fourth percentile, exceeded by only 16 per cent of all Mexican families.

By 1960, inflation, falling cotton and wheat prices, and continued population growth had had a marked effect. The 1960 annual income of 879,000 pesos was equal to an average monthly income for the 252 San Miguel families of 34.7 1937 pesos, 231 1956 pesos, and 244 1957 pesos. Compared with the income of all Mexican families in 1956, San Miguel's average family income was exceeded by 76 per cent of all Mexican families, according to the Cline data. According to Navarette's figures, San Miguel's relative position was even lower—at the tenth percentile, with 90 per cent of Mexico's families having a larger income. Though there is a considerable discrepancy in these two estimates, both point to a decline from an economic position in the top half of the Mexican population to one close to the bottom. The total difference between the 1950 percentile rank calculated on the basis of ejidatario families only and the 1960 rank calculated on the basis of all San Miguel families is 58 percentile points according to Cline's study, and 74 points according to Navarette's. In both cases, approximately half of the difference is attributable to the decrease in real income of the ejido, and half to the increase in the number of families.

Estimates of the minimum income required to provide food and clothing for a Mexican family were also made by these authors. Navarette's estimate for 1957 was 700 pesos (1960: 151); by this figure, 59 per cent of the Mexican families in her study did not earn an income adequate to meet the minimum requirements to sustain health. Cline (1962: 116) believed that 300 pesos was sufficient in 1956. The federal government's minimum wage of 537 pesos per month provides a third basis for estimating a family's essential income. In 1950 San Miguel's per-family income of 592 1957 pesos exceeded two of these minimum standards; by 1960, however, its per-family income of 244 1957 pesos was below all three estimates. Per-ejidatario family income was higher than the lowest of the three estimates, but there were few ejidatario families in San Miguel that were not supporting the family of a libre son or other relative.

Making matters worse for San Miguel families was the uneven distribution of wages over the year. In the years when there were profits, credit for food and clothing could be obtained, but from 1957 to 1967 neither the ejidatarios nor the libres had profits after the crops were sold with which to pay off credit extended during the year. The ejido receives 75 per cent of its wages in the spring and summer months; in the fall and winter, when the need for clothing and fuel is greatest, only 25 per cent comes in, an average of 131 pesos per month per ejidatario (Table 29).

TABLE 29
Total Monthly Wages Advanced to San Miguel in 1960 for Work in Cotton and Wheat
(*Pesos*)

Month	Amount	Month	Amount	Month	Amount
Jan.	22,400	May	13,900	Sept.	7,700
Feb.	17,200	June	113,600	Oct.	6,300
Mar.	30,200	July	84,800	Nov.	14,400
Apr.	31,300	Aug.	66,700	Dec.	13,100

SOURCE: Ejido Bank's Plan of Operations: loans for wages.
NOTE: The total amount actually advanced for wages according to San Miguel records (Table 25) was 502,000 pesos. Either that figure included wages advanced for other kinds of work, or else the bank's planned loan for cotton and wheat was some 80,000 pesos less than the actual loan.

In the absence of detailed data, it is difficult to pinpoint all the causes of San Miguel's decreasing income and to determine the precise role of each factor in the overall decline. It is obvious that the major factors are the general inflation and the stationary cotton prices; but other elements are involved as well. The production costs, including wages, of San Miguel have not increased year by year at the same rate as Mexican wholesale and consumer prices; rather, they have stayed on plateaus and then risen suddenly in a single year, which points to changes in operating procedures or in Ejido Bank policies (or possibly in bookkeeping methods). Repairs of the cotton gin, now over 20 years old, have become a larger and larger item; by 1960 machinery repairs were 8.5 per cent of the ejido's total non-labor budget, and much of this was the repair of the cotton gin. Another major item whose cost has increased is electricity for the well pumps; in 1960 this came to 222,000 pesos, slightly over a fifth of the non-labor expenses. A primary reason for this enormous electric bill, almost U.S. $1,000 a month, is the steadily dropping level of the region's water table, which is now below the level of many pumps. Those wells that still have water require more pumping and hence more electricity. In addition, electric power in Mexico was nationalized between 1960 and 1964, and the government's new standard rates were higher than the rates previously charged in the Laguna region and other areas with a good supply of electricity.

The Libres. Between 1936 and 1953 the number of libre families increased as the sons of ejidatarios grew up and married and as libre sons-in-law came to San Miguel, usually to live with their in-laws. With the declining ejido income, the ejidatarios were less able to support additional libre families, and libres began to leave San Miguel. Census figures indicate that between 1952 and 1960 the emigration rate equaled the rate of natural increase. From time to time the federal government has raised the hopes of the libres that they might become ejidatarios or colonists in new irrigation projects (such as the ill-fated Papaloapan project in Vera Cruz), but no libres from San Miguel, to my knowledge, ever benefited from these. Similarly, the Central Union had not succeeded in its attempts to have the government create new ejidos in the Laguna region from the estates that have more than the legal maximum of 150 hectares, with title nominally in the hands of various relatives.

The libres who have remained in San Miguel have benefited from the Ejido Bank's policy of increased wages for both collective work and individual parcela work; but they were already doing most of the collective work in 1953. The income of the libres is not enough to support a family by even the most minimal standards. The median income of 40–49 pesos they earned during the 16-week period studied in 1953 was only one-thirtieth of the lowest of the minimal family standards (300 pesos per month); only four libres, during those 16 weeks, had earned as much as 20 pesos per week, less than a third of the minimum standard. Even the few libres fortunate enough to work regularly on private farms could earn no more than 165 pesos per month, still only about half of the monthly family minimum. It is clear that San Miguel's libres have never earned enough to support their families. They live with their ejidatario relatives and make themselves useful in whatever ways they can—in parcelas, in collective work, in household chores, and in any other ways the ejidatarios suggest. With the ejidatarios themselves facing the situation of a declining income from 1950 on, the position of the libres has become even more precarious, and they have begun to leave the ejido in increasing numbers, as, indeed, have a number of ejidatarios. Among those who have left have been some of the most capable workers in the ejido, both ejidatario and libre, and especially those who were fairly certain to find work in the urban world.

Education. In 1953 San Miguel's elementary schoolteachers did not encourage the ejido children to go on to high school or vocational school. The school was an educational dead end except for a few children whose parents had the motivation and financial resources to send them on to the public elementary schools of Matamoros and Torreón, and to the higher vocational schools of the region.

When I returned to San Miguel in 1966, there was a new group of teachers with a much more realistic conception of the function of education and a much more positive attitude toward the collective ejidos and rural workers. The principal of the school, himself the son of an ejidatario and educated in nearby Matamoros, knew many of the ejidatarios personally. He identified with and was sympathetic to the collective ejidos and their problems, and was anxious to provide the ejido children with an education that would fit them for the economic opportunities available to them.

With a complete elementary school of six grades and a staff of teachers aware of the community's needs, the children of San Miguel now have an opportunity for an education that can lead to increased economic opportunity and social mobility in the larger urban society.

In 1953 there was a drop-out rate of more than 50 per cent in each grade, with only one out of seven children completing even the four grades available in San Miguel; by 1966 most of the children were completing all six grades in San Miguel and over a dozen boys had obtained government scholarships to study agriculture and bookkeeping in the regional vocational schools.

San Miguel's teachers participate in regularly scheduled district meetings, where they are informed of new Ministry of Education policies and encouraged to enforce old ones. I attended one of these all-day meetings at Matamoros in 1966, a pleasant affair in which the teachers freely complained about paperwork and talked to the supervisors in a way that indicated considerable autonomy and good morale in spite of the restrictions of the bureaucratic organization.

Though the federal government must be given due credit for providing money for teachers' salaries, the ejido has provided strong economic support to the school from the outset. Not many ejidos have a six-grade elementary school; that San Miguel does is evidence of the community's positive attitude toward education and of a strong ejido organization that takes advantage of opportunities for community projects offered by the government. In fact, San Miguel measures up exceptionally well in the educational opportunities it provides its children in comparison with other parts of the nation. Half of the rural schoolchildren in Mexico have no schools to attend at all, according to Tannenbaum (1963: 107), and the vast majority of the existing rural schools have only three grades. Nor is the educational problem confined to the rural population. In Mexico City, for instance, an estimated 50,000 to 100,000 children, of about one million, do not have access to a school. The federal government, which prides itself on a relatively large education budget, simply cannot keep pace with the increasing population, rural or urban.

Division and Conflict in San Miguel

The most dramatic change in San Miguel between 1953 and 1966 was not the community's economic decline or the emigration of its libres and ejidatarios, but its division into two bitterly opposed groups, one of which wanted to form its own separate ejido credit society. The Separatistas, as the members of this group were called by the others, contended that from 1954 on, the ejido officials had permitted Ejido Bank officials to cheat San Miguel, and perhaps had even been guilty of graft themselves. This was, of course, the same period that had seen an enormous rise in the costs of cotton production because of inflation and the increased electricity expense; these factors alone could easily have accounted totally or in large part for the lack of profits year after year. Nevertheless, the suspicions of the Separatistas were supported by the fact that a number of Ejido Bank officials, including several of the bank presidents, had been accused and convicted of large-scale graft, and by the fact that in the purchase of seeds, insecticides, and equipment, as well as in the processing and marketing of cotton, Ejido Bank officials had ample opportunity to cheat the ejido, with or without the collusion of ejido officials. An authority on Mexico's banking system, Charles Anderson, has described the problem of dishonesty in agrarian banks like the Ejido Bank as follows:

Those banks which deal in a high volume of small transactions, which serve an economically and socially disadvantaged clientele, and which operate through a large number of small agencies have been the target of criticism for failures ranging from excessive red tape to outright theft, graft, and larceny on the part of bank officials.

The agrarian banks have borne the brunt of this criticism. The form of organization adopted by these institutions, in which credit cooperatives were to distribute credit granted by local bank agencies to individual clients, has not generally been successful. Since the great majority of members of these cooperatives are unprepared to understand or to cope with the responsibilities of cooperative membership, these organizations have generally been an ineffective way both of administering credit and of organizing an interest group which could demand honest, efficient performance by bank personnel. Unscrupulous bank agents and cooperative leaders may conspire to defraud both the clients and the bank itself. Attempts by the central bank office to insure honest performance through detailed rules and regulations and a centralization of decision-making have more often resulted in binding the functions of the bank in a maze of red tape than in correcting administrative ills.

(Anderson and Glade 1963: 141)

Though Anderson's description supports the possibility that the Separatistas' allegations of graft were true, it is not necessary to assume this to be the case in order to understand the role of the Ejido Bank in the Separatista movement in San Miguel. The crucial points are (1) that the Separatistas believed they could do better on their own, and (2) that the Ejido Bank permitted or even encouraged them to form their own credit society, thus dividing the ejido into two separately functioning sectors.

After the Cárdenas era, the bank took the position that it was only a financial institution, not a social welfare agency, and should operate according to the institutional banking criteria of minimizing financial risk and maximizing profits. That this was indeed the policy in the Laguna region was clear in the division of most of the ejidos, with the blessing and encouragement of the Ejido Bank, with loans going only to the groups that were considered good financial risks. The split in San Miguel occurred somewhat later than most. But frustrations had built up after eight profitless years, and they culminated in 1962 in a petition by 25 ejidatarios to form a new sector. The petition was bitterly rejected by both the San Miguel officials and the Central Union on two grounds: that some of the Separatistas no longer worked or even lived in San Miguel, and that the ejido would be seriously weakened. But there were other, unstated reasons for the refusal to countenance a new sector. First, the breaking away of the Separatistas from the ejido could have led to a loss of credit for the remaining ejido group, particularly if the new sector were more successful economically. Second, the creation of sectors would have been a departure from the idea of collective ownership of property, to which the Central Union was ideologically committed. Third, there was the composition of the separatist group itself; it included many former ejido officials, among them some of the most knowledgeable and urbanized ejidatarios in San Miguel. There was some feeling that these men placed their own interests above those of the community. Finally, the conflicting groups represented different national political parties and political views. The Separatistas referred to the main group as Reds because of their loyalty to the Central Union; they in turn were called the Whites. Almost certainly, if the Separatistas had been successful, they would have joined the progovernment Confederación Nacional de Campesinos, the CNC. It should be noted, however, that the shift in allegiance on the part of the former ejido leaders was not out of concern for the Marxist orientation of the Central Union: when I was in San Miguel in 1953 these same officials had regarded such political considerations as irrelevant. The ejido belonged to the Central Union simply because it was effective in dealing with the Ejido Bank.

The denial of the petition in 1962 marked the beginning of a period of open conflict. Over the next four years, continued appeals were sent to the Agrarian Department to settle the issue, but to no avail; the department was either unable or unwilling to make a decision. Meanwhile, the disputants harassed one another and argued in the ejido assembly. Considering the gravity of the problem, it is a tribute to their patience and moderation that physical violence did not erupt until 1967. In January of that year, while I was in San Miguel, the main group of ejidatarios blocked the conduits that carried water to the houses of the Separatistas, who retaliated by having the ejido officials arrested. Even then the federal government failed to step in to resolve a conflict that was certain to lead eventually to violence. Only a month later it came. On February 9, 1967, 20 of the Separatistas armed themselves with guns. Proceeding into the fields before dawn, they marked off an area that they claimed as theirs and settled down to defend it. This act was essentially an admission of defeat. Nevertheless, the ejido president, Perfecto Morales, went to the fields to investigate. He and his son were killed, and six other ejidatarios were wounded in a last futile gesture by the separatist faction.

The Agrarian Department was thus relieved of the problem of deciding whether to grant the petition for a new sector, for 14 of the 29 Separatistas left San Miguel. The 15 who remained in the community lost their ejidatario rights; they are now libres and have given up the attempt to form a separate sector. But the victory of the main group was only a single incident in the continuing conflict between the Ejido Bank and the Central Union over the question of whether agricultural credit should be based primarily on the needs of the ejidatarios or on the institutional criteria of traditional banks.

San Miguel is one of the few ejidos in the Laguna region that is not divided into two sectors. However, the Ejido Bank continues to control San Miguel's agricultural policies and economic destiny almost as tightly as the land company controlled the economic destiny of the peons of San Miguel. The important difference is that the ejido has a political voice, and so may influence the Mexican government. The failure of the Agrarian Department to act on the Separatistas' petition to become a new sector, as well as the unwillingness or the inability of any other federal agency, including the Ejido Bank, to support them, is evidence of the political power of the collective ejidos and of the point of view represented by the Central Union. The ejidatarios have enough strength to resist direct attacks on the collective ejido system and, as in the case of San Miguel, can even resist the implementation of policies that they strongly oppose.

6
San Miguel and Other Collective Ejidos

The Laguna Ejidos

In 1953 there were approximately 700 collective ejidos in Mexico, about one-fourth of them in the Laguna region, and the rest distributed throughout Mexico, as shown in Table 30. Between 1953 and 1960 there was apparently a decline in the total number of collective ejidos, though it is difficult to determine to what extent; since many ejidos were divided into sectors, the total number of collective units reported by the Ejido Bank in fact increased. A questionnaire sent to all of the regional branches of the Ejido Bank showed that 129 ejidos listed by ten branch banks as collective in 1953 were no longer so classified in 1960. Whether this indicated a real change in the economic organization of these ejidos or merely a change in their classification by bank officials is not clear. The last is a possibility, for both Eckstein (1966) and I found instances of government officials minimizing the number and importance of collective ejidos. One official even informed me that there were no longer any collective ejidos in Mexico, since each ejidatario now had his own parcela.

On the assumption that the 129 ejidos reclassified between 1953 and 1960 were, in fact, no longer collective, and that the apparent increase in the number of collective ejidos in the Laguna region and the Yaqui-Mayo area were only increases in the number of collective units, then the estimated decrease between 1953 and 1960 was approximately 25 per cent. Still, it seems likely that the decrease was considerably less than this estimate: in the Laguna

TABLE 30

Number of Collective and Individual Ejidos in Mexico, 1953, by Location of Bank Branch

Ejido bank, location	Number of ejidos in area[a]	Individual[b]	Collective[c]
Torreón, Coahuila	355	43	268
Ciudad Obregón, Sonora	203	100	74
Veracruz, Veracruz	377	309	67
Córdoba, Veracruz	288	189	44
Apatzingán, Michoacán	65	15	38
Tepic, Nayarit	227	175	36
Mérida, Yucatan	217	162	33
Zacatecas, Zacatecas	147	116	24
Celaya, Guanajuato	559	498	18
Toluca, Mexico	412	342	11
Ciudad Victoria, Tamaulipas	354	312	11
Brisenas, Michoacán	131	109	10
San Luis Potosí, San Luis Potosí	252	213	10
Tapachula, Chiapas	171	148	9
Pachuca, Hidalgo	200	209	9
Aguascalientes, Aguascalientes	165	135	8
Other areas	3,360	2,972	15
Total	7,483	6,047	685

SOURCE: Eckstein 1966: 170.
[a] Includes 751 ejidos (10%) that did not respond to the questionnaire on which the data in the other columns are based.
[b] Includes 158 individual ejidos that have form of producer cooperative.
[c] Includes 189 collective ejidos that divided their land into parcelas and 486 that did not.

region the number of ejidatarios operating with the assistance of the Ejido Bank increased from 20,573 in 1957 to 25,340 in 1964 (90 per cent of whom were in collective ejidos).

The effects of the collective ejido economy has differed from place to place, depending on the nature of the community prior to the creation of the ejido. Fortunately for our understanding of these effects, developments in several of the regions with different cultural and economic traditions have been studied and provide some basis of comparison with the ejido of San Miguel.

Three studies of the Laguna region ejidos are Clarence Senior's *Land Reform and Democracy*, which is based primarily on regional statistics obtained prior to 1940; a 1966 study by Salomon Eckstein, *El Ejido Colectivo en Mexico*, which analyzes agricultural census data for ten regions of Mexico, four of them compris-

Other Collective Ejidos 151

ing the Laguna region; and an intensive economic study of the ejidos created out of the Tlahualilo Land Company published in 1964 by Juan Ballasteros Porta. I have used some of Senior's data earlier in this study, and shall not discuss his work here. The Eckstein book studies the relative economic efficiency of collective ejidos, individual ejidos, and private farms in the Laguna region, as measured by per capita income, aggregate productivity, and returns per unit of land, labor, and capital. Eckstein finds that in the higher income regions, where there is enough good land and enough water to justify the use of machinery and other large-scale farming methods, the collective ejidos are economically more efficient than the individual ejidos and compare favorably with both large and small private farms. He finds no evidence in these regions that underemployment was made more severe by a higher degree of mechanization. Conversely, he finds that in the low-income regions, where no benefits are obtained from mechanization, the individual ejidos are superior to the collective ejidos in all respects, though inferior to the private farms. After a certain optimum scale of operation has been reached, he has concluded, low income ejidos do not benefit from collective organization. Their only gain is some saving of labor, which in view of underemployment is not beneficial.

Though Eckstein's concern is primarily economic, he also discusses the social organization of the collective ejidos and their relation to the federal government. He, too, attributes the process of dissension and division within the ejidos primarily to the federal government, specifically to the effort made by the Ejido Bank to undermine the Central Union and to switch the allegiance of the ejidatarios to the rival CNC. It was Eckstein's contention in 1959, at the beginning of the presidential administration of López Mateos, that the government promised to be more favorably disposed to the collective ejidos than former administrations; but that promise remained unfulfilled.

Ballasteros Porta's study of the ejidos of Tlahualilo shows a development that closely parallels San Miguel's. As in San Miguel, there were more ejidatarios in the Tlahualilo ejidos at the outset than the number of peon workers in the hacienda, which led to problems not only of overpopulation but of work methods and income distribution. But in spite of these difficulties, the ejidos were successful in the first few years, thanks to government help and favorable conditions. Shortly after 1940, however, the government withdrew its support of the ejidos, calling them a Marxist experiment, and after 1942 the Tlahualilo collective ejidos disintegrated. A major contributing factor in this process was widespread corruption, in which both Ejido Bank officials and ejido leaders were involved.

Though the production of cotton, wheat, and corn had increased markedly after 1936, with the yields increasing threefold in the case of cotton, the Tlahualilo ejidos had seen the same enormous population growth as San Miguel. In just under 25 years, the adult male population almost doubled, increasing from 1,305 in 1936 to 2,576 in 1960. At the time Ballasteros Porta made his study, 60 per cent of the work of the Tlahualilo men was done outside the ejido.

Ballasteros Porta proposes several changes in the ejido organization in the Tlahualilo area. Among other things, he recommends that the ejidatarios be relocated so that each of those remaining will have ten hectares of irrigated land; that ejidatario rights be taken from those who are renting their land rather than working it themselves; that the sectors be abolished and independent credit societies be organized for each ejido wherever necessary, taking into account the available resources and the number of ejidatarios; that the work be reorganized and a rational system of distribution of profits in proportion to the amount and quality of work be instituted; that the ejidatarios be educated to cooperative principles; that agricultural credit be made available in sufficient quantity, and on time; and that every ejido be encouraged to provide its own working capital.

It is clear from these recommendations that the Tlahualilo ejidos faced many of the same problems as San Miguel. What they come down to simply are too many people for the economic resources, and inadequate assistance from the Ejido Bank. But in

Other Collective Ejidos

one important respect the Tlahualilo ejidos are more typical of the Laguna region than is San Miguel: The ejido of San Miguel incorporated an entire hacienda, whereas 13 ejidos were carved out of the Tlahualilo hacienda, leaving the best land, the water, and the capital resources in the hands of private owners. Thus, one of the main purposes of the collective ejidos, to maintain the integrity of large-scale operations, was not accomplished.

Other Ejidos

Among the studies of ejidos outside the Laguna region is that of J. Hernández Segura, of the National School of Agriculture of Mexico, who examined the ejidos of the state of Michoacán in the late 1950's. In that area two large and well-organized estates, Lombardia and Nueva Italia, were sold to the Ejido Bank in 1938 for two million pesos and were divided into nine ejidos, which promised to become show places that would demonstrate the success of collective operations; by the time Segura made his study they had completely disintegrated as collectives.

In the first stage of their existence, between 1938 and 1944, the nine ejidos were operated as a single cooperative enterprise. But after a few profitable years, they faced heavy losses, the result of a combination of poor harvests and agricultural techniques, bad administration of Ejido Bank loans, and overinvestment in community projects (including a stadium costing 250,000 pesos). Between 1944 and 1952, the ejidos splintered; by 1956 all the land had been distributed among the ejidatarios to be worked individually, and in 1958 the cattle herds were divided. Of 55 Nueva Italia ejidatarios interviewed at random by Segura in 1957, 51 per cent cultivated their own land, 14 per cent did not cultivate any land at all, 22 per cent rented to outsiders, and 13 per cent had some kind of sharecropping arrangement in which the owner worked his own land and was paid a wage, preferring this to the risks and responsibilities of independent farming. In these ejidos the Ejido Bank recognized the de facto renting of ejido land and actually withheld the rents due the landlord.

Hernández concludes that Nueva Italia succumbed under the

weight of unemployment. It had been hoped that additional work would be supplied by diversifying the community's activities, but the ejido received no guidance or support in this endeavor from the Ejido Bank. In the end, despite the promise of the land, the failure of the Neuva Italia collective ejidos was a failure of organization and management; the attempt to diversify foundered on the Ejido Bank's policies, which permitted overextension of debt.

The most detailed anthropological study of collective ejidos is Charles Erasmus's study of the Yaqui-Mayo area of Sonora. This study, part of a larger work entitled *Man Takes Control,* concentrates on how the ejido organization has affected the social life of the Yaqui and Mayo Indian communities.

The area Erasmus studies is quite different from the Laguna region. It has not reached the limits of its natural resources, either in land or in water, and the inhabitants of the ejido villages are Indian peasants, not a rural proletariat, as in the Laguna region. The Yaqui and Mayo ejidos also differ from the Laguna ejidos in many respects. For example, the average land holding in the Yaqui region is about 24 hectares per ejidatario. Another major difference is that in the Mayo area an ejido is not in itself a community but an association that cross-cuts community lines; the members of a single ejido credit society may come from as many as ten villages. A third important distinction is that the transition from rural ejidatario to urban worker has been hindered in the Yaqui-Mayo area by cultural differences between the Indians and the other inhabitants, a problem that does not exist in the Laguna region. Finally, no social class lines have developed within the Yaqui-Mayo ejidos, in spite of economic differences.

Erasmus finds that half of the ejido land in the Mayo area is rented or sharecropped. According to him, the Yaqui Indians do not like the collective ejido system and look forward to the promised division of ejido lands, this in spite of their long battle to preserve their collectively owned reservation lands. Moreover, the Indians have effectively circumvented collectivization with the active help of the Ejido Bank. It is Erasmus's view, however, that if the Indian ejidatarios were permitted to sell their holdings, the

Other Collective Ejidos

land would rapidly be consolidated in the hands of a few persons. He cites the case of an Indian estate that was divided among 27 heirs, each of whom inherited about 80 hectares. Within a few years, one-fourth of the heirs had sold all their land, another fourth rented their holdings, and though half still worked the land, their inheritance had dwindled to an average of 7.2 hectares; only four heirs had kept their original 80 hectares. "If allowed to do so, Erasmus contends "most ejidatarios would sell their land."

Land ownership and management tends to shift from Indian to Yeri (non-Indian) hands whenever the law permits, and it would seem that this shift is toward more efficient and profitable management and greater national production. The land reform laws that attempt to keep the use-ownership and management of farms fixed form a policy to protect the Indian (ejidatario) but not necessarily the economy at large.

(Erasmus 1961: 235)

Though Erasmus's focus is somewhat different from mine, it is clear that the same processes are at work in the Yaqui-Mayo region as in San Miguel, and that the same relationship obtains between the ejidatarios and the federal government.

Relatively little attention has been given to the other collective ejido areas of Mexico. The Yucatan henequen area, for example, has not been studied by any anthropologist or sociologist. In any case, in that area the government and the private landowners have reached an agreement on the ownership of property and equipment and on the administration of henequen plantations that in effect makes the ejidatarios wage workers; and the communities there are not really collective ejidos, for they have virtually no economic autonomy. Of the other collectives in Mexico, a cattle ejido at Cananea, which was organized in 1958, seemed to Eckstein to be operating successfully in 1963, thanks in part to technical advice from the University of Arizona and the active support of high officials of the Mexican Agricultural Department. And one source (Chonchol 1957) contends there were some prosperous collective ejidos still operating in Sinaloa in 1957, though the Ejido Bank maintains there are no collective societies in that area now.

The Mexican government's Center for Agrarian Studies is making intensive economic and sociological studies of the effects of the whole agrarian reform since the Revolution, including the collective ejidos. When published, these studies will be the first major attempt to collect and synthesize data on all the collective ejidos of Mexico.

General Conclusions

Even in the absence of studies that would make a detailed comparison of San Miguel's experience and the general ejido history possible, certain generalizations can be made on the basis of the available data.

1. Collective ejidos have been successful in terms of agricultural productivity. Their crop yields have increased since 1936 and compare favorably with those of private farms.

2. The regions where collective farming constitutes a major part of the agricultural economy, notably the Laguna and Yaqui-Mayo areas, have prospered and grown since 1936. Two probable causes for this economic growth are the development of an internal market as the purchasing power of families grew, and the increased capital expenditures of the private farms, which intensified their operations.

3. The net real income of the collective farms has not kept pace with their increased production, in large part because of the increased expenses of production, a relative decrease in the value of agricultural products, inflation, and the inadequacies of the Ejido Bank policies.

4. Most collective ejidos have faced the same major problems, overpopulation and underemployment, though not always to the same degree as San Miguel.

5. With the government's encouragement and assistance (and pressure), the collective ejidos were divided into individual parcelos between 1942 and 1945. Though the most important decisions and agricultural operations continued to be collective activities, this move facilitated further decollectivization, first expressed in various individual adaptations, which were permitted though

Other Collective Ejidos

not officially approved by the Ejido Bank and the Agrarian Department (such as the use of libre labor and the renting or sharecropping of parcelas), and then in the division of ejidos into sectors or separate credit societies.

6. In all of the studies of collective ejidos, the Ejido Bank appears as one of the primary causes of both economic and political difficulties. The bank is seen to have failed in three ways: it provides too little credit, too late; its technical assistance programs are inadequate; and some of its officials have been guilty of, or have permitted, graft.

7. The promise of the collective ejidos with respect to noneconomic aspects of community life—health, education, family relationships, religion, and leisure time activities—has not been fulfilled. With the exception of education and health, there has been no appreciable qualitative or quantitative improvement in community activities since the changes that took place as soon as the ejidos were created. (According to the agricultural censuses of 1940 and 1950, the proportion of ejidos, collective and individual, having schools, water, and medical services, actually decreased.)

8. Although the ejidos have not been as economically or politically successful as they might have been with continued government support, they have provided an opportunity for some ejidatarios and their children to escape the cycle of poverty of the hacienda peon. Some ejidatarios have used their income as a base from which to seek other economic opportunities, and the children of some have gone beyond the primary school level to vocational high schools.

9. The collective ejidos constitute a political force that must be reckoned with by the government. As long as the ejidatarios prosper they have a stake in the economic and political status quo, which they will fight to maintain. Politicians and government officials who oppose the collective ejidos seldom do so openly or directly. Rather, they count on the slow decline of the ejidos through a combination of inefficiency, graft, and governmental indifference.

7
Alternative Courses for the Collective Ejidos

Future Paths

It is difficult to become involved in the life of a community like San Miguel without wondering what its future will be. Because of the many variables involved, one cannot predict with any certainty what path the ejido system will follow. But this study of San Miguel would be incomplete without a discussion of the possible courses that the government and the collective ejidos could take in the next decade. With respect to the relationship between the federal government and the collective ejidos, there seem to be four major possibilities. The ejidos could be incorporated into the government, with federal ownership and operation of the land, and with the ejidatarios becoming salaried workers, as in the case of the oil industry and the railroads. Or, the relationship could be one of total dependence, in which the ejido is not directly controlled by the government, but is dependent on and under the close control of the Ejido Bank—in short, the present relationship. A third possibility is the development of a relationship of interdependence, with the government encouraging and facilitating gradual steps toward greater independence in credit, technical assistance, and new sources of income. Finally, the ejidos could break with the government altogether, which would mean no government controls, but also no government support other than that afforded private agriculture. These alternatives do not exist merely as abstract logical possibilities. They are deliberate choices to be made, decisions that can only be taken in and through the specific circumstances of Mexico's political and economic realities. To make discussion of these alternatives meaningful, therefore, it is first useful to take a brief look at Mexico's post-Revolutionary political course.

Post-Revolutionary Mexico may be divided politically and economically into three periods: the period from 1925 to 1934, in which the nation was ruled by a coalition of regional military strong men; the Cárdenas period from 1935 to 1940, in which the power of the military declined and the guarantees and goals of the 1917 Constitution, which favored the rural and urban workers, were seriously implemented through ejido grants, the strengthening of trade unions, and the nationalization of basic industries, including the railroads, the oil industry, and the telegraph system; and the period from 1940 to 1970, in which the five post-Cárdenas administrations have promoted rapid industrialization at the expense of the living standards of both rural and urban workers.

Under the Constitution of 1917, Mexico has a republican form of government with elected legislatures and executive officials at local, state, and federal levels. In practice, the election of legislatures, state governors, and even local mayors is merely the ratification of the president's selection. The president appoints every important political official, including his own successor; his power is limited only by the duration of his six-year term of office (without the possibility of reelection) and by an oligarchy of 20 to 30 ex-presidents, generals, and wealthy men, whose approval he needs in selecting his successor. The presidents of Mexico are invariably selected from this oligarchy, which includes every shade of political opinion from the far left to the far right; former President Cárdenas was, until his death in November 1970, one of the most powerful members of this group.

The policies of the administrations since Cárdenas have been directed primarily toward increasing the total productive capacity and income of the country. The benefits to workers, who constitute most of the population, have been not in increased income but rather in improved education, health facilities, and other social services. The results of the government labor policies have been summarized as follows:

> Since Cardenas left the presidency in 1940, agricultural labor, especially the two million families in communal agriculture and the million families on privately owned minifundia plots, has received credit and some price supports, but only a political demagogue can contend that the preponderant majority of these three million families are much better off than they were twenty years ago. The same goes for the trade unionists; brute industrialization, forced capital accumulation, and attempts to balance international payments have left little room for higher real wages. The immediate objective [of the government] is to industrialize and commercialize the nation; the redistribution of income and the placing of greater purchasing power in the hands of labor can come later.
>
> (Brandenburg 1964: 16)

From 1940 to 1946 President Ávila Camacho's administration continued many of the programs begun under Cárdenas, but with Mexico's entry into World War II, social welfare took second place to the war effort. President Alemán (1946–52) gave priority to rapid industrialization and large government projects, and adopted policies that favored the wealthy and powerful. Ruiz Cortínes (1952–58), López Mateos (1950–64), and Díaz Ordaz (1964–70) also pursued a policy of rapid industrialization, but with more moderation than Alemán and with programs more acceptable to the Cardenistas. Since 1940 private agriculture has been given priority over ejido agriculture, and the primary emphasis has been on increased production (especially of cash export crops) and the building of dams to irrigate new croplands. Under Alemán, the maximum size of private farms was increased and titles of "inaffectibility" were issued to private farms to ensure against future expropriation for ejido grants.

The three administrations that have directed Mexico since 1952 have not favored the workers as the Cárdenas administration did, but neither have they gone to the extremes of favoring industry, commerce, and the rich and powerful, as the earlier Alemán administration did. The controlling oligarchy and the official party both contain a wide spectrum of political opinion, and both Cardenistas and Alemanistas are to be found at all levels of government and in all administrations. At times the government has opposed and jailed leaders of both right-wing and left-wing groups, but in general there is considerable freedom of speech.

Certain political policies and ideas have come to be so accepted by the Mexican people that it would be political suicide to openly oppose them. Primary among these is the principle that Mexico's resources belong to and must be used for the benefit of the Mexican people. One corollary of this theme is economic nationalism. Mexico is especially sensitive about United States economic control, and its attempts to attract American capital since 1940 have been accompanied by safeguards against American control. A second corollary is that the benefits of the Revolution and of economic development must go to all the Mexican people. When a president's policies become explicitly or obviously opposed to this notion of social and economic justice, as did Alemán's, he is in political trouble.

The government is the steward of the nation's resources on behalf of all the people. Private ownership of land, industry, and commerce is permissible and even desirable as long as it clearly contributes to the national welfare, but most Mexicans feel that government control and ownership of industry, commerce, transportation, and agriculture, best carries out this policy. Some economists believe there is an inherent conflict between productive efficiency and government control, and it is true that certain government-controlled enterprises have had severe problems; but some of these industries, notably the railroads and the electric companies, were in economic trouble before the government took them over. The primary difference seems to be not over the question of the relative merits of government and private ownership, but over the question of priorities: is the emphasis to be on capital growth and increased productivity or is it to be on raising the per-capita income and the standard of living of the workers? In general, the government gives more attention to meeting the immediate needs of the workers than do private employers. Nevertheless, since 1940 it has stressed economic production and national development over a more equitable distribution of the nation's income.

One of the basic tensions in the Mexican political system is the fact that Cárdenas's philosophy, which favored the working people, is supported by the majority of Mexicans but by only a minority of the oligarchy responsible for filling the presidential seat. If the present emphasis on industrialization, production, and economic growth does not at the same time provide more jobs, genuine wage increases, and improved living standards for the majority of Mexicans, there is a chance members of the Revolutionary Left could return to power. In that event, the government's priorities would shift, leading to further government ownership and control of industry, higher wages and higher taxes, price controls, increased support of the ejidos and consumer cooperatives, a strengthening of the trade unions, decreased reliance on foreign private capital, and a general slowing up of industrialization. At present, however, this is only a possibility. It is too early yet to predict in what direction Mexico will go in the next six years, under its new president, Luis Echeverría.

San Miguel's relationship to the federal government has been one of almost total dependence since it was created in 1936. During the Cárdenas regime, the government was supportive, paternalistic, and concerned with the total life of the community. Since Cárdenas, the Ejido Bank has been concerned almost exclusively with the economic aspect of the community; its policies have aimed at ensuring the repayment of loans, and little else.

During the entire Cárdenas and post-Cárdenas period, the federal government has offered no realistic alternative to total dependence. Total independence and autonomy are legally possible, but the government has offered this only on terms of immediate and total loss of all government support. The ejidos and ejido sectors that have been cut off from Ejido Bank credit, technical assistance, marketing, and other services have inevitably turned to subsistence agriculture by individual ejidatarios. Few ejidos have voluntarily chosen this drastic path to economic autonomy.

Alternative 1: Continuation of Total Dependence. In the next decade, the relationship between the collective ejidos and the federal government will, most probably, be a continuation of the total economic dependence that has existed since 1936. This means that loans for the weekly wages will continue to be made by the Ejido Bank, now merged with the Agricultural Bank (Banco Agricola), which formerly gave credit only to private farmers. More than likely, as a result of this merger, the collective ejidos will have to meet more stringent conditions to obtain loans. The chances that the ejidos' debts will be cancelled, in whole or in part, in a bad crop year or a low market situation also seem diminished. Dependence on government bank loans will continue to mean that the Ejido Bank will determine what crops will be produced, as well as the details of the agricultural methods used. Payment of wages will remain conditional on the completion of a specific amount of work.

As long as the Ejido Bank continues to be San Miguel's sole source of credit, the ejido will have to conform to the bank's wishes in the matter of crop diversification, an area in which the bank officials seem to have shown a complete lack of imagination. Virtually every study of the Laguna region has concluded that agriculture there should become diversified to include dairy and poultry farms, vineyards, and fruit and vegetable production. This is the direction that the private farms of the Laguna region have taken. The government, however, has not encouraged the ejidos to grow anything but cotton and wheat. Though a government agricultural experiment station at Matamoros is available to San Miguel, its recommendations cannot be followed by the collective ejidos without the permission of the Ejido Bank. San Miguel's first step toward crop diversification was in the cultivation of grapes, and the experiment seems to be successful. But new crops or livestock require additional credit and adequate technical assistance; until an ejido can obtain both, diversification is impossible.

Unless there is a fundamental change in policy, the government can be expected to do little to create new jobs or new sources of income for the libres and ejidatarios. Up to the present, it has acted only when starvation has seemed imminent in the Laguna region. But the use of public works projects (most often the cleaning of irrigation canals) as relief measures has proved inadequate, even for a temporary period. Army conscription has also provided income and employment for young men, but Mexico has decreased its military expenditures, and the government no longer requires a large army to maintain political control. As for the creation of new sources of income, the history of the government's main intervention in this connection in San Miguel is not encouraging: San Miguel installed its cotton gin in spite of the Ejido Bank, not because of it.

Libres and ejidatarios have found work (without government assistance) on private farms in the Laguna region, in other parts of Mexico, and in the United States; others have found jobs in Torreón, and a few have developed small businesses in the ejido. The limits of outside employment seem to have been reached, however. The only contribution the government has made in this area is the vocational training of young men and women for skilled specialist work. But few young men in San Miguel have obtained government scholarships, and those that have will not return to the ejido to work; education is a way to get out of the ejido, not a way to improve one's life in it. The ejido schools, which have the potential for increasing the ejidatarios' prospects, are neither designed for nor prepared to accept this role.

In summary, the outlook is dismal. Most probably, the relationship between the collective ejidos and the federal government will continue to be one of total economic dependence, in which the government will offer little or no initiative for innovation, and the ejido will have to overcome the inertia and red tape of the Ejido Bank in order to institute even small adaptive changes.

Still, we should at least look at the other possibilities.

Alternative 2: Total and Immediate Separation. By law, the collective ejidos have the option of severing all ties with the federal government; those that have done so have reverted to subsistence agriculture. Collective ejidos can obtain credit from private banks or companies, and have sometimes done so, but private lenders are not always willing to make loans, and when they do, they tend to charge what the traffic will bear. Collective ejidos have turned to private lenders only in desperate situations or for relatively small loans.

To be sure, successful ejidos could put aside enough profits to finance their crop for the coming year, freeing themselves of their dependence on the Ejido Bank; but at the same time they would become totally responsible for any losses. Moreover, if an ejido does not obtain its loans from the Ejido Bank, it forgoes the bank's technical assistance which, whatever its shortcomings, has been virtually the only technical aid available. Without it, most ejidatarios would fall back on subsistence crops like corn, beans, and chili, rather than cash crops requiring intensive, highly capitalized production methods. This has happened in a number of ejidos and ejido sectors that have been denied credit by the Ejido Bank. The collective ejidos are almost certain not to exercise this option: to choose total and immediate separation from the government is to choose a road to economic inefficiency and lowered living standards.

Alternative 3: State Ownership and Operation. Opponents of the collective ejidos have contended that ejidatarios are, in effect, peons on state-owned haciendas. This was clearly not the case in San Miguel, where the income of the ejidatarios, at least until 1956, was appreciably higher than it was before 1936, and where the inalienability of ejido lands gave a degree of security that the peons never had. For the ejidos and the ejido sectors unable to obtain Ejido Bank loans, however, direct government ownership and operation might be preferable to the present system.

The Mexican government has increasingly taken over industries it considers important in the national economy, making the workers in these industries government employees. There is no reason to suppose such a course would prove less effective in agricultural production than it has been in industrial production. Under the present system, it is the ejidatarios who bear the economic loss in the event of failure, even though they may have had little control over the factors leading to that failure. If the government operated a collective ejido at the request of the ejidatarios, the workers would at least be assured of a guaranteed wage, and at best a share of the profits. Moreover, the government might come to realize the importance of the total life of each ejido community, rather than limiting its concern to agricultural production.

Apart from the question of desirability, it is unlikely that there will be any move in this direction in the foreseeable future. The collective ejidos, with their total dependence on the Ejido Bank, produce the cash crops the government wants produced, and have kept pace with private agriculture in yields. Economic difficulties of any kind, whether owing to bad weather, low market prices, or inflationary increases in production costs, are borne by the ejidatarios. As things stand, the government can play the role of Good Samaritan, taking credit for whatever it does to alleviate the situation. Were the ejidos to become state-owned enterprises the government would then be responsible for paying its "employees" a living wage, and for any losses that the ejidatarios incur. Fortunately for the government, most ejidatarios would probably oppose state ownership of the collective ejidos, since the land is their one inalienable possession.

Alternative 4: Interdependence. The federal government has offered the collective ejidos a choice between the extremes of total dependence and total independence. But to be successful, independence would have to be achieved gradually, progressing through a series of intermediate steps. Government assistance leading gradually to more and more ejido independence would be a feasible alternative and would alleviate many of the existing problems. Government officials have often stated that their policies are aimed at increasing ejido autonomy, but seldom has their rhetoric been accompanied by action. In my judgment, to be effective, any government program designed to increase ejido autonomy would have to meet the following seven conditions.

In granting credit, the government would have to be concerned with the long-range economic success of the ejido, not merely with

Alternative Courses 167

the community's loan payment for the current year. When poor weather, lack of water, low market prices, or other conditions beyond the control of the ejido created a deficit, the government would have to grant subsidies, much as it does in support of both private and state industries. This would ensure that advances toward ejido autonomy and self-sufficiency would not be wiped out by a single bad year.

The collective ejidos' suggestions on credit would have to be given serious consideration. A case in point: the ejidos' proposal that wages be increased and spread more evenly through the year to permit better budgeting of family income.

Technical assistance—for crop diversification, improved farming methods, marketing, and the like—would have to be made available to the collective ejidos, whatever their relationship with the Ejido Bank. The government's agricultural experiment stations are a step in this direction; under the proposed program, ejidos would be allowed to decide for themselves how to use the information and advice of the experiment stations. If the permission of the Ejido Bank were no longer a requisite, the collective ejidos could try various kinds of crops and livestock, as the private farms of the Laguna region have, and as agricultural experts have long advocated.

The government would have to provide technical assistance and loans for the creation of light industries and handicrafts in the ejidos. Sanford Mosk's *Industrial Revolution in Mexico* (1950) concludes that small, decentralized industries are better suited to the resources of Mexico and its needs than heavy, centralized industry. In San Miguel the construction of the cotton gin has done much to alleviate the chronic unemployment. Most collective ejidos could themselves do much more in this direction, and the government could initiate and finance handicraft industries in both ejido and non-ejido communities at a minimal cost. Cotton-weaving and the production of articles made of cotton cloth are an obvious possibility in the Laguna region. Studies could establish what specific handicrafts and products would be most successful and should be promoted.

The government ought to give technical assistance and loans to family and community self-help projects. Housing, potable water, and facilities for health, education, and recreation are needed in every ejido. With government aid, ejido families could be employed to improve their own houses and community facilities. Dr. Isabel Kelly, in her survey of housing in the Laguna region in 1953, found great need for such a program. She found also that some ejidatarios had come to rely so completely on the government they would not improve their own houses unless they were paid in cash for their labor.

The government would have to permit and encourage flexible adaptation of the ejido charters to legalize such arrangements as land rental and use of libre labor. Changes in these vital areas have been either made illegally or established by bureaucratic fiat. Open discussion of such basic changes and participation of the ejidatarios in the decision-making process would be important steps toward ejido independence.

Finally the government would have to be concerned with the total life of the collective ejidos, not merely with their economy, and with the living conditions of the libres as well as the ejidatarios. The Ejido Medical Service, for example, would be available to libre and ejidatario families without distinction, thus dealing with the ejido as a total community, not merely as an economic organization from which the libres are excluded.

One of the best-documented studies of the Laguna collective ejidos is *La Comarca Lagunera* (published in 1940 by the semi-official League of Socialist Agronomists), whose list of contributors reads like a who's who of government agricultural officials. Among other things, they recommended what I refer to here as interdependence.

The Ejido Bank has taken over the administration and control of agriculture in the ejidos, and has not always put these functions in the hands of competent personnel. Thus the participation of the ejidatarios in the management of the ejidos has been scorned, and because of this the Ejido Bank has been rightly called the new proprietor of the Laguna region. As soon as the collective ejidos have been correctly organized, with land distributed among the ejidatarios in a reasonable way, the ejidatarios should manage their own affairs, for in the last two years they have shown that they have the ability required to carry out commercial dealings and administration. This is especially noticeable because the Ejido Bank has entrusted technical administration to professionals who do not always have administrative ability and in many cases know nothing of how to work with the common people. (Liga 1940: 483)

There are no doubt many ways the government could encourage small steps toward greater independence. I have emphasized only those I consider the most important. The point is: total and immediate separation of the ejidos from governmental technical and financial assistance does not constitute independence. Ejido autonomy must be achieved slowly and incrementally. Moreover, in a complex industrial nation like Mexico, autonomy will not mean total separation from government assistance; the relationship will always be one of interdependence.

Mexican government officials have often said that their goal for the collective ejidos was the kind of interdependence we have just discussed. Why then have the Ejido Bank and other government agencies not adopted policies to achieve that goal? Why have they, on the contrary, prolonged the total dependence of the collective ejidos? There are no simple answers to this question. But a plausible explanation can be found when one examines the role the collective ejidos play in the overall economic and political plans of the government.

From the standpoint of government planners, some of the most important economic and agricultural functions of the collective ejidos are the efficient production of cash crops for export; the efficient production of food and industrial crops for the industries of Mexico and the urban population; the maximum repayment of loans to the Ejido Bank; and the purchase of fertilizers, electricity, oil, gasoline, machinery, and other items from state-controlled industries at rates fixed by the government. These four economic and agricultural objectives are achieved as well—and perhaps even better—by policies that keep the collective ejidos totally dependent on the government as they would be by policies fostering increased interdependence.

The political functions of the collective ejidos during the Cárdenas administration were to support the administration's policies, which were generally more favorable to workers and peasants than those of any previous or later administrations, to serve as a model for increased worker participation in the control of industry and agriculture, and to give the rural workers a larger share of the nation's total income by reducing the income and power of the large landowners. Since that time, the political role assigned the collective ejidos has been quite different. They are intended to keep a large number of rural workers politically quiescent and under the control of the Ejido Bank and the government-controlled CNC. Further, by their lack of success, they are intended to demonstrate that there should be no further extension of the ejido program, either in number or in land area and, more generally, that the Cárdenas policies of worker participation in the control of industry and agriculture do not work.

The collective ejidos were originally designed to support the health, education, family life, and other social programs of each ejido community. But the post-Cárdenas governments, opposing the basic ideas underlying the collective ejidos, separated these social functions from the ejido economic organization. The shift in the collective ejidos' political functions since 1940, along with the abandonment of social functions, is not consistent with a policy that would promote ejido independence. The political and economic priorities of the Mexican government since 1940 have been on industrialization, on increased capital growth, and on the promise of the eventual trickling down of economic benefits to the rural and urban workers. These priorities are consistent with the continued policy of total dependence.

From the standpoint of the collective ejidos, the preferred relationship with the government is a relationship of interdependence. A government policy to this end would undoubtedly be expensive; it would not, however, cost as much as the present subsidization of rapid industrialization. It is unlikely that the new Echevarria administration will favor interdependence over continued total dependence; but if it should become very secure (or very insecure), such an alternative is possible. Many Mexicans have strongly ad-

Alternative Courses 171

vocated a more positive federal policy toward the collective ejidos, but no administration has yet adopted such a policy.

Social Alternatives

A second group of alternatives involves the relationship between the libres and the ejidatarios. However modified by kinship ties, the primary relationship between these two groups is economic, rooted in the ejidatarios' collective ownership of the community's total productive capacity and the exclusion of the libres from all rights in the ejido economy.

Though the ejido economic organization is for all practical purposes the de facto political and community organization of San Miguel, lack of political autonomy in some ways binds the ejidatarios to the libres, notably in their second-class citizenship in the municipio of Matamoros, which prevents or makes difficult the assumption of responsibility and initiative in the organization of noneconomic community activities, as well as economic activities in which libres have equal rights.

In regard to the basic libre-ejidatario relationships, I see at least four alternative courses as possibilities for the future: a continuation of the present situation, in which libres are not participating members of the ejido economic organization and in which neither libres nor ejidatarios are organized into a politically recognized community; the granting of ejidatario status to all or most of the libres in the community; the attainment of political status by the community of San Miguel, making both libres and ejidatarios full citizens with a vote and a voice in all activities involving the total community; and the simultaneous achievement of economic and political equality by libres and ejidatarios, that is, the simultaneous attainment of the second and third alternatives. Any of these alternatives would have more chance of success if accompanied by attention to ways of slowing down the rate of population increase.

Alternative 1: Continuation of the Present Situation. The number of libres in San Miguel now far exceeds the number of ejidatarios, a situation that obtains in other collective ejidos as well.

As a result, the democratic community envisioned in the ejido charter is being transformed into a two-class community. Though the primary social role and status of most libres is that of an ejidatario's son, the libres also constitute a lower status group without economic rights in the community; they are permitted to reside in the community only because of their kinship to the ejidatarios. In general, the government and the ejido tend to act as if the libres are not going to stay in the community permanently, but are there only temporarily until new ejido lands become available or until employment opportunities develop outside the ejido. Despite the uncertainty and vagueness of such opportunities in the future, the belief that they will develop is used as a justification for not incorporating the libres into full participation in community life: there is no reason to restructure the community for the sake of temporary residents.

For my part, I see nothing to suggest that the present course will not continue, that libres will not continue to reside in the community and to work when convenient for the ejidatarios, with only as much participation in community matters as the ejidatarios deem appropriate—in short, a community consisting of a privileged minority and a dependent, subordinate majority.

Alternative 2: Granting of Ejidatario Status to Libres. Since San Miguel is allowed only a fixed quantity of irrigation water under the current system of water distribution, granting ejidatario status to a libre would decrease the share of irrigated land and income of each of the present ejidatarios. Even if ejidatario status were to be granted to a few libres as ejidatarios leave the community or are expelled, only a small fraction of the libres of San Miguel would be affected. Not until the ejido's water supply is increased or new sources of work and income are created could any significant proportion of the libres be given ejidatario rights without decreasing the income of the present ejidatarios. The Central Union has been fighting, without success, for the distribution of the water of the Laguna region to the ejidos on a first-priority basis, in accordance with government regulations issued at the time the ejidos were created. The government not only has failed to adhere strictly to the letter of the law in this connection; it has permitted private landowners to increase their proportion of the water supply of the region over the years by issuing them permits for new wells.

In brief, a significant proportion of libres could be granted ejidatario rights only if the water distribution system of the Laguna region were changed to give the ejidos most of the water that now goes to private landowners, or if new sources of work and income were created by the government. Both of these developments are technologically possible but are unlikely to take place, at least in the next decade, unless the federal government radically changes its attitude toward the collective ejidos.

Alternative 3: Attainment of Full Political Status. The low status of the libres is more than a matter of exclusion from the economic organization of the ejido; it is compounded by the fact that neither they nor the ejidatarios have any legally recognized rights as citizens of the community of San Miguel. Residence in the ejido and participation in ejido activities are by consent of the only legally recognized organization in the community, the Ejido Credit Society. The hacienda land company may be gone, but the community of San Miguel is still limited: its *raison d'être* is not the total life of all its inhabitants but only the economic production and income of the ejidatarios.

It may in fact be too late for San Miguel and other collective ejidos similarly divided into conflicting groups and separate credit societies to achieve genuine community solidarity. The economic conflicts that have divided the collective ejidos are deep, and they are reflected in other aspects of community life. But assuming that these conflicts may eventually diminish, is it possible that the collective ejidos could become genuine communities with an organization reflecting concern for all aspects of social life? And, more specifically, could the collective ejidos become the kind of democratic community envisioned in their charters, with every worker having an equal right and opportunity to participate in the basic decisions that affect his life?

The numerous kinship ties in San Miguel, the proximity of

one residence to another, and the short distance of the ejido from Torreón and from other ejidos make it difficult to conceive of this community as ever being merely an economic organization. When San Miguel was a hacienda and its people simply employees of the land company, most of the kinds of noneconomic activities that bind a community together were not possible, but even then there were kinship ties and religious fiestas as well as recreation, friendship, and social interaction patterns that were not fixed by the demands of economic production. In that period, San Miguel was exclusively an economic organization in the sense that all other activities were subordinated to economic production. Unless the economic organization of the ejido were to be controlled by some outside group or individual, it is inevitable that interaction in economic activities will be accompanied by many noneconomic activities. San Miguel and ejidos like it can never be purely economic organizations. Whatever the position of the Ejido Bank or other agencies, government economic activities affect every aspect of the total community.

If San Miguel were recognized as a political entity, made up of all the persons who have resided there for some time and who participate in the community's activities, nonmembership in the Ejido Credit Society would not prevent the libres from participating in noneconomic matters or in certain economic decisions. A shift in this direction would spur the creation of new sources of income, such as handicrafts or cottage industries, that would benefit ejidatarios as much as libres. But several obstacles would have to be overcome if San Miguel were to become a self-governing unit. For one thing, though the conception of the ejido as a democratic community is understood and accepted by the ejidatarios, it would not be easy for them to give up their superior position over the libres in noneconomic community activities. More important, the federal government would have to recognize the community as a legal political entity, with the power to choose its own elected officials and full political rights for the libre population. The prospects of this happening are not good, even if a Cardenista government were to come to power again; centralized political control

Alternative Courses

is one of the characteristic features of Mexican government. In any case, granting San Miguel some degree of political autonomy and legal recognition would not automatically make the libres an integral part of the community. It would, however, make them citizens of a community and not merely poor relatives living on the property of an economic corporation that is owned by their kinsmen.

Alternative 4: Simultaneous Achievement of Political and Economic Equality. As we have seen, it is unlikely that San Miguel (or other collective ejidos) will either grant ejidatario status to its libres or become politically recognized communities in which both libres and ejidatarios have the right to organize and to participate in the noneconomic activities of the community. But these are the goals the creators of the collective ejidos hoped would be attained eventually.

If either development took place, the chances of the other taking place would be immeasurably increased. If sufficient work and income were created to grant ejidatario status to all the libres, the present ejidatarios would not have reason to exclude the libres from full citizenship in the political community. Similarly, if the libres were citizens of the community, with the right to live there and the right to participate in noneconomic activities, granting them ejidatario status would not be as big a step. Notwithstanding the intentions of the planners of the 1930's, this alternative is perhaps the least likely of all to occur. It could not be successfully achieved by the ejidos without strong government support, both financial and political.

Alternative 5: Slowing Population Growth. The increase in the libre population, which has so changed the economic and social life of the ejido, has been due to three factors. In order of importance, these are: a decreased infant and child mortality, a continued high birth rate, and a net in-migration of libres. If by some unlikely combination of circumstances every libre in San Miguel and the other collective ejidos were given ejidatario status today, the present birth and death rates would lead to a recurrence of the same problems now confronting the ejidos within 25 years.

Emigration out of the ejido, which has acted as the primary population check in San Miguel since 1935, has not kept pace with in-migration. Unless jobs can be created for migrants, the act of migration alone cannot help the overall Mexican situation. Migration across the border as bracero labor, once an important if temporary source of work for Mexican farm workers, has been all but shut off by the United States in the past decade. And the industrial development into which the government has put most of its resources has not created a significant number of new jobs for rural workers.

The possibility of checking the population growth by decreasing the birth rate runs into strong obstacles in Mexico. In urban industrial countries, the middle class, especially persons who are upward mobile with expectations of a higher standard of living, have limited their family size because of a desire for increased economic and material well-being, both for themselves and for their children. In Mexico, birth control, practiced primarily among the upper and middle classes, is in conflict with the teaching of the Catholic Church, regardless of the level of society. A doctor who had been with the Ejido Medical Service for years told me that it was impossible to discuss birth control publicly in Mexico because of the climate of opinion. Still, since the Mexican government is not officially allied with the Church, and since most of the ejidatario men are indifferent, if not hostile, to the Church, it seems possible that some kind of birth control might be accepted by ejido families.

If the Ejido Medical Service were to institute a program of planned parenthood in conjunction with its maternal health care programs, it is quite possible that ejido families would accept family planning. Government support of the Medical Service in this respect is as likely under moderate and rightist administrations as under a Cárdenas-type administration, and it seems possible that one or two determined women in each ejido might institute such a program if they could obtain the help of the Ejido Medical Service.

It is clear after three decades that though there are advantages

Alternative Courses

to collective action and planning, there are also urgent needs for individualization and for adaptations to change demographic and economic conditions, adaptations that are not possible within the existing ejido charters and government regulations.

Alternatives to Pressures for Decollectivization

What are the alternatives open to the collective ejidos in face of pressures for decollectivization, which have already led to widespread land rental, the use of libre hired labor, and the de facto transfer of ejidatario rights to substitutes?

The primary advantages of collective action and planning are well known. In connection with the collective ejidos specifically, they include economies effected in large-scale operations by the use of machinery; economies effected in purchasing, marketing, and bookkeeping; easier access to credit and better guarantees of loan repayment; and the inalienability of the land by government fiat. To some extent, however, these advantages are offset by many problems, some of them not susceptible to legal solution as the ejido charters are now written. These problems include the lack of sufficient work in the ejidos to provide full employment for the ejidatarios; the presence of large numbers of libres willing to exchange their labor for little more than the opportunity to continue living in the community; the rapidly increasing population; inflation and the rising costs of production; and dissension caused in part by economic difficulties and in part by political differences that reflect political forces outside the ejidos.

The alternative courses that are possible in response to the dual needs for collective planning and work on the one hand, and for flexible adaptation to these problems on the other, seem to me to be as follows: a return to the original conception of completely collective ejidos, with no variations permitted; a continuation of the present situation in which individual deviations and adaptations are permitted by the nonenforcement of laws and regulations prohibiting land rental and the use of hired labor, and by ad hoc bureaucratic decisions made to suit the convenience of the Ejido Bank and other government agencies; the creation of legal modifications based on the tendencies toward land rental, the use of libre labor, and the increasing employment of ejidatarios outside the ejidos; the partition of the collective ejidos into individual ejido grants, with the government encouraging and facilitating the formation of voluntary producer cooperatives; and the partition of the collective ejidos into individual ejido grants, with each ejidatario operating completely on his own without government assistance.

Alternative 1: Completely Collectivized Ejidos. A return to the collective ejido as it existed in the first several years of the ejido program would be possible only if the government decided to take over the ejidos and to operate them as a nationalized industry. The various moves toward decollectivization could not be reversed without such drastic government action. In that event, the government would have to assume responsibility for the libres, to create work for them or to expel them from the ejidos. There is nothing to suggest the government is willing to assume any such responsibility—for either the libres or the ejidatarios.

Alternative 2: Continuation of the Present Situation. This seems to me the most probable course in the future. Though the government has not permitted legal changes in the charters of the collective ejidos, it does not enforce the laws prohibiting the rental of land and the use of libre labor. In addition, the Ejido Bank has brought about various kinds of changes, by bureaucratic ruling, including the parcelization of the collective ejido lands and the division of ejidos into sectors. Such ad hoc bureaucratic decisions, together with the nonenforcement of laws, have not only contributed to conflict and dissension within the ejidos but deprived the ejidatarios of the opportunity and the right to participate in decisions that vitally affect them.

Alternative 3: Legal Modifications. The charters of the collective ejidos provide, among other things, that ownership of ejido land is vested collectively in the ejido assembly, which is to decide how the land shall be used, how the work shall be performed, and so forth; that individual ejidatarios have the right to participate equally in the decisions of the ejido assembly but cannot transfer their ejido membership to another person; and that an ejidatario must himself participate in the work of the ejido in order to share in the profits—he cannot hire others to do his work for him.

The most important adaptations of these basic provisions have been the creation of sectors, the practice of having libres do the work of their ejidatario relatives, and the rental of land. The use of libre labor and the rental of land may be attributed to many factors, but both Erasmus (1961) and Eckstein (1966) attribute these practices primarily to "economies of scale"; that is, when individual parcelas are not large enough to keep an ejidatario fully employed, it is more efficient for one man to work in more than one parcela, thus releasing other men for work outside the ejido. This basic demographic and economic fact conflicts with the present ejido charters and with government regulations. The modifications that have been made to permit the rental of land and the use of libre labor are inadequate in that they have not been made openly, legally, and with the active participation of the ejido assemblies.

If the ejidatarios could legally change their charters and the government regulations, they would (in all likelihood) legally permit libres to work for an ejidatario with certain restrictions to prevent their exploitation, ejidatarios to rent their parcelas to other ejidatarios with restrictions to ensure that the renter works the land himself, and ejidatarios to sell their membership in the ejido either to a libre, subject to the approval of the ejido assembly, or to the ejido. By the adoption of these rules, an ejidatario could leave the ejido to look for other work without being penalized for doing so. He might take two or three years to try out a new job, renting his land during that period. If the job worked out, he could sell his ejido membership; if not, he could return to the ejido and resume full participation. Payment to the ejidatario could be made by the ejido assembly over a period of several years or by the libre who obtained the ejido membership. Possible abuses of such a system could be fairly easily controlled. For example, limits could be set so that an ejidatario who rents his land —say, four years in a row, or six years out of ten—would be compelled to sell his ejido membership. Joint control of the rental of land and the sale of ejido membership could be vested in the ejido assembly and the government to ensure that ejido membership is transferred only to libres who are qualified and competent. The government could require that the number of ejidatarios in an ejido not be permitted to fall below a given number, and make any other conditions that seem necessary to ensure the viability of the economic organization.

With this kind of flexibility, the collective ejidos would become in effect regulated stock corporations, with the workers owning all the stock. There would be increased incentive and opportunity for the ejido to buy land and wells from private landowners and to increase the number of ejidatarios in the ejido. The notion that the collective ejidos would be weakened if individual ejidatarios could rent or sell their land is based on the erroneous assumption that the only alternative to the present collective land tenure is private, individual landownership. There are many ways that individuals can become members of an economic organization, participate in it, and leave it, intermediate between the two extremes of absolute collective ownership and absolute individual ownership. The alternative proposed here, that the collective ejidos become stock companies, regulated by the government in order to protect individual ejidatarios and to prevent the breakup of the ejidos, is one of these intermediate solutions. By this proposal, an ejidatario might sell his membership or give it up, but if he did, a libre would take his place and the ejido would continue to provide work and income for the same number of rural families. Other intermediate solutions are also possible, requiring only that the government make a serious effort to create strong, independent, productive ejidos.

Since the 1940's the federal government has stated its opposition to the dogmatic and rigid application of "collectivist" ideas in the ejidos. Yet the policies that have been substituted for those of the Cárdenas administration have been ad hoc policies and bureaucratic fiats, which have not dealt openly and honestly with the realities of overpopulation and underemployment in the collective ejidos. Until individual ejidatarios can legally rent their land, use libre labor, and sell their ejido rights in some way similar to the program I have outlined, these activities can be expected to go on illegally and to be a continuing source of conflict within the ejidos.

Alternatives 4 and 5: Partition of the Ejidos. Some might propose that the processes of decollectivization in the ejidos could in the end lead to the individual type of ejido. In that event, the ejidatarios could find it difficult to grow cotton or other cash crops requiring large amounts of capital, and would probably revert to the cultivation of corn, beans, and other subsistence crops. Since such a drastic change would be a severe blow to the nation's economy, it is almost inconceivable that the federal government would permit individualization on a grand scale, though it has already occurred in many of the ejido sectors that cannot get credit from the Ejido Bank.

It is, of course, possible that individual ejidatarios could grow cash crops such as cotton by forming voluntary producer cooperatives. The success of such ventures is problematical, however, and would require considerable government support. By and large, it would be much more difficult for the government to create and maintain voluntary cooperatives than it would be to reform and modify the present collective ejidos along the lines proposed above.

The alternatives available to the collective ejidos depend in part on national economic, political, and demographic trends that are beyond their control. But workable alternatives do exist— alternatives that depend as much on a clear conception of the problem as they do on additional economic support by the government. In my opinion, the alternatives most likely to succeed are those that make use of the experience and competence of the ejidatarios themselves, and that take into account their ability to participate in decision-making. In instances where the government has trusted the ejidatarios' judgment and has worked with them to arrive at joint decisions, the results have generally been good. This is in sharp contrast to many situations in which the government, for whatever motive, has excluded the ejidatarios from the decision-making processes. The result has often been economic inefficiency, graft, and political conflict. No conception of economic organization or of community development can be successful that is imposed on a community. It must be initiated and continually modified in accordance with changes in the situation and the changing perceptions and experiences of the people involved. The failures of the collective ejidos have been due, in large part, to the failure of the government to follow this basic principle of community development. The successes of the collective ejidos have been due, in large part, to the application of the principle of democratic participation, and to the dogged effort and spirit of the ejidatarios themselves, whose lives are staked on the success of this vast experiment in economic and political democracy.

References

Adams, Richard. 1964. "Rural Labor," in John J. Johnson, ed., *Continuity and Change in Latin America*. Stanford Univ. Press, Stanford, Calif.

Anderson, Charles, and William P. Glade. 1963. *The Political Economy of Mexico*. Univ. of Wisconsin Press, Madison.

Avila, Manuel. 1969. *Tradition and Growth: A Study of Four Mexican Villages*. Univ. of Chicago Press, Chicago.

Ballasteros Porta, Juan. 1964. *¿Explotación individual o colectiva? El caso de los ejidos de Tlahualilo*. Instituto Mexicano de Investigaciones Económicas, Mexico, D.F.

Banco Nacional de Crédito Ejidal. 1945. *El sistema de producción colectiva en los ejidos del Valle del Yaqui, Sonora*. Mexico, D.F.

Beals, Ralph L. 1932. *The Comparative Ethnology of Northern Mexico Before 1750*. Univ. of California Press, Berkeley.

Benedict, Ruth. 1946. *Patterns of Culture*. Penguin Books, New York.

Brandenburg, Frank. 1964. *The Making of Modern Mexico*. Prentice-Hall, Englewood Cliffs, N.J.

Chevalier, François. 1963. *Land and Society in Colonial Mexico*. Univ. of California Press, Berkeley.

Chonchol, Jacques. 1957. *Los Distritos de Riego del Noroeste*. Centro de Investigaciones Agrarias. Mexico, D.F.

Cline, Howard. 1962. *Mexico: Revolution to Evolution, 1940–1960*. Oxford Univ. Press, London.

Eckstein, Salomon. 1966. *El ejido colectivo en México*. Fondo de Cultura Económica. Mexico, D.F.

Erasmus, Charles J. 1961. *Man Takes Control: Cultural Development and American Aid*. Univ. of Minnesota Press, Minneapolis.

184 References

Feder, Ernest. 1970. "La función social de la tierra," *El Trimestre Económico*, XXXVII, 145.
Fernández y Fernández, Ramón. N.d. *Notas sobre la reforma agraria Mexicana* (Serie Monografías Num. 2). Escuela Nacional de Agricultura, Chapingo, Mexico.
Fernández y Fernández, Ramón, and Ricardo Acosta. 1961. *Política Agrícola*. Fondo de Cultura Económica, Mexico, D.F.
———. 1961. "La colectiva ha muerte. ¡Viva la colectiva!," Chapingo, I, 3.
Freithaler, William. 1968. *Mexico's Foreign Trade and Economic Development*. Praeger, New York.
Guerra Cepeda, Roberto. 1939. *El ejido colectivizado en la Comarca Lagunera*. Banco Nacional de Crédito Ejidal, Mexico, D.F.
Hernández Segura, J. 1959. Estudio de las condiciones económico-agrícolas de las Sociedades de Neuva Italia. Escuela Nacional de Agricultura, Chapingo, Mexico.
Kelly, Isabel. 1955. *Survey of Housing in the Ejido of Cuije*. Servicios Medicos Rurales, Torreón. Mimeo.
———. 1965. *Folk Practices in North Mexico: Birth Customs, Folk Medicine, and Spiritualism in the Laguna Zone*. Univ. of Texas Press, Austin.
Lewis, Oscar. 1951. *Life in a Mexican Village: Tepoztlán Restudied*. Univ. of Illinois Press, Urbana.
———. 1957. "Mexico Since Cardenas," in Council on Foreign Relations, ed., *Social Change in Latin America*. Alfred Knopf, New York.
Liga de Agrónomos Socialistas. 1940. *La Comarca Lagunera*. Talleres de Industrial Gráfica, S.A., Mexico, D.F.
McBride, George. 1923. *The Land Systems of Mexico*. Am. Geog. Soc., New York.
Mendieta y Núñez, Lucio. 1964. *El problema agrario de México*. Editorial Porrua, S.A., Mexico, D.F.
Mendizabal, Miguel Othon de. 1946. "El problema agrario de la Laguna," in Vol. IV of *Obras completas de Miguel Othon de Mendizabal*. Mexico, D.F.
Moreno, Pablo C. 1951. *Torreón: biografía de la más joven de las ciudades Mexicanas*. Talleres Gráficos "COAHUILA," Saltillo, Mexico.
Mosk, Sanford. 1950. *Industrial Revolution in Mexico*. Univ. of California Press, Berkeley.
Navarrete, Alfredo. 1964. "Comparative Analysis of Policy Instruments in Mexico," in Werner Baer and Isaac Kerstenetsky, eds., *Inflation and Growth in Latin America*. Richard D. Irwin, Inc., Homewood, Ill.
Navarrete, Ifigenia M. de. 1960. "Income Distribution in Mexico," re-

References

printed in *Mexico's Recent Economic Development: The Mexican View*. Institute of Latin American Studies. 1967. Univ. of Texas Press, Austin.

Nuevo código agrario, 1943. Asociación de Empresas Industriales y Comerciales, Mexico, D.F.

Paz, Octavio. 1961. *The Labyrinth of Solitude: Life and Thought in Mexico*. Grove Press, New York.

Pimentel, José Reyes. 1937. *Despertar Lagunero*. Talleres Gráficos de la Nación, Mexico, D.F.

Poleman, Thomas T. 1964. *The Papaloapan Project*. Stanford Univ. Press, Stanford, Calif.

Redfield, Robert. 1955. *The Little Community*. Univ. of Chicago Press, Chicago.

———. 1956. *Peasant Society and Culture*. Univ. of Chicago Press, Chicago.

Schiller, Otto. 1961. "Explotación agrícola colectiva en México," *Chapingo*, I, 4.

Scott, Robert. 1959. *Mexican Government in Transition*. Univ. of Illinois Press, Urbana.

Secretaría de Recursos Hidráulicos. 1951. *Estudio agrológico detallado del distrito de Riego en la región Lagunera*. Distribuidora de Algodones, S.A., Mexico, D.F.

Senior, Clarence. 1958. *Land Reform and Democracy*. Univ. of Florida Press, Gainesville.

Shafer, Robert J. 1966. *Mexico: Mutual Adjustment Planning*. Syracuse Univ. Press, Syracuse, N.Y.

Simpson, Eyler. 1937. *The Ejido: Mexico's Way Out*. Univ. of North Carolina Press, Chapel Hill.

Simpson, Lesley Byrd. 1952. *Many Mexicos*. Univ. of California Press, Berkeley.

Singer, Morris. 1969. *Growth, Equality, and the Mexican Experience*. Univ. of Texas Press, Austin.

Stavenhagen, Rodolfo. 1968. "Seven Fallacies About Latin America," in James Petras, ed., *Latin America: Reform or Revolution?* Fawcett, New York.

Steward, Julian. 1959. "Perspectives on the Plantation," in *Plantation Systems of the New World*. Pan-American Union, Washington, D.C.

Tannenbaum, Frank. 1929. *The Mexican Agrarian Revolution*. Macmillan, New York.

———. 1950. *Mexico: the Struggle for Peace and Bread*. Alfred Knopf, New York.

———. 1963. *Ten Keys to Latin America*. Alfred Knopf, New York.
Tomasek, Robert, ed. 1966. *Latin American Politics*. Doubleday, Garden City, N.Y.
United Nations Statistical Yearbook. 1954, 1958, 1961, 1968. Lake Success, N.Y.
United Nations Yearbook of International Trade Statistics (for 1967). 1969, Lake Success, N.Y.
Vernon, Raymond. 1963. *The Dilemma of Mexico's Development*. Harvard Univ. Press, Cambridge, Mass.
———, ed. 1964. *Public Policy and Private Enterprise in Mexico*. Harvard Univ. Press, Cambridge, Mass.
Whetten, Nathan L. 1948. *Rural Mexico*. Univ. of Chicago Press, Chicago.
Wolf, Eric. 1959. *Sons of the Shaking Earth*. Univ. of Chicago Press, Chicago.
———. 1966. *Peasants*. Prentice-Hall, Englewood Cliffs, N.J.

Index

Agrarian Code, 41, 46, 69
Agrarian Committee, 30, 34, 96f
Agrarian Department, *see* Mexico
Agrarian Laws (1915), 17, 19
Agricultural Bank (Banco Agricola), 163
Agriculture, xi, xii, 9–13 *passim*, 69f, 134, 162–65 *passim*. *See also* Alfalfa; Corn; Cotton; Wheat
Aguanaval River, xiv, 3ff, 9, 11, 26, 54
Aguayo, Marqués de, 10, 11
Alemán, Miguel, 52, 160
Alfalfa, 69–70
Amusements, *see* Baseball; Leisure activities; Social activities
Anderson, Charles, 145–46
Ávila Camacho, Manuel, 50, 160

Bailes (dances), 111–12
Ballasteros Porta, Juan, 151f
Banks, agrarian, 146. *See also* Ejido Bank
Baseball, 27, 110–11
Birth control, 34, 176
Birth rate, 34, 175f
Businesses, 73–75

Cananea, Sonora, 155
Cárdenas, Lázaro, 18–19, 20, 54, 109, 159, 162
Cardenistas, 160–61
Catholic Church, 16, 107–8, 176. *See also* Religion
Center for Agrarian Studies, 156
Central Union of Collective Ejidos, xiv, xvi, 8, 53, 108, 121, 172; and Ejido Bank, 50–51, 52, 148, 151; and Separatistas, 146–47
Children, 16–17, 34–35, 101–2, 108–9, 122–24, 143–45
Climate, 3–4
Collective system: benefits of, 20, 68, 177; modified, 56–57; attitudes toward, 70–71, 154; and libres, 71, 86, 87–88
Comarca Lagunera, La, 168–69
Communications, 7–8, 113–14
Compadrazgo (godparenthood), 124–25
CONASUPO cooperative, 131
Confederación Nacional de Campesinos (CNC), xvi, 50–51, 147, 151, 170
Constitution (1917), xi, 100; of ejido, 38, 93, 178–79
Cooperatives, 31, 60f, 96, 122, 131, 181
Corn, 14, 69–70, 71
Cortez, Hernando, 106
Cotton, 4, 14, 63, 72, 75; prices, 18, 77–78, 135; collective labor in, 64–65, 66; yields, 71, 76–77
Cotton gin, 27, 30, 60–61
Credit, 84, 148, 163, 165ff
Crops, *see* Alfalfa; Corn; Cotton; Wheat

Dance of the Indians, 105–6
Decision-making, 40, 45, 49, 182. *See also* Ejido assembly
Decollectivization, 40, 57, 65–73 *passim*, 134, 153, 156–57, 177–81 *passim*

Index

Díaz Ordaz, Gustavo, 160
Diet, 14, 60, 70, 101
Doctrina (Sunday school), 108–9
Drinking, 115–16

Echeverría, Luis, 162, 170–71
Eckstein, Salomon, 150–51, 155, 179
Economic nationalism, 161
Economic policies, national, 159–62, 166, 169f
Education, 16–17, 27–30, 99–103 *passim*, 143–45, 157, 164
Ejido assembly, 30, 40f, 48–50, 65, 92, 95
Ejido Bank, xvi, 61, 65, 67, 72, 146; power of, 41, 43, 70, 148, 163, 168–69, 178; and Central Union, 50f, 151; and wages, 63–64, 84, 85–86, 137, 143; corruption in, 145–46, 152; criticized, 153f, 157, 162
Ejido constitution, 38, 93, 178–79
Ejido Credit Society, 173f
Ejido Medical Service, 25, 35, 50, 79, 95, 168, 176
Ejido system, establishment of, xi–xii, 17–23 *passim*, 38
Erasmus, Charles, 154–55, 179
Exchange labor, 71
Executive committee of ejido, 41–42. *See also* Official positions
Expropriation, 20f, 58

Faena (labor as a civic duty), 107
Federal Agrarian Code, 41, 46, 69
Federal government, xiv–xv, 37–38, 99f; relationship to ejidos, 50, 151, 158, 162f, 180–81, 182; and water rights, 54, 55–56; economic policies of, 159–62, 166, 169
Federal Water Commission, 6, 12
Fiestas, 103, 104–5, 106, 110
Food, 14, 60, 70, 101

Gerente (cotton gin manager), 46
Gómez Palacio, Durango, 6
Government, ejido, 40, 95. *See also* Ejido assembly
Government, federal. See Federal government; Mexico
Government, state, 19, 52

Hacienda system, 9–17 *passim*

Health care, *see* Ejido Medical Service
Hernández Segura, J., 153
Housing, 30, 31–32, 59, 168

Illegitimacy, 118
Income, 71, 75, 82–87 *passim*, 121f, 156; per capita, 81; decline in, 134–42 *passim*; sources of, 164
Inheritance, 39f
Indians, 9–10, 154–55

Jiménez, Juan Ignacio, 11
Joint families, 15–16, 89, 126
Joint work, 126–27
Juárez, Benito, 11
Juez (official position), 52–53

Kelly, Isabel, 168
Kinship, 16, 97, 124–27, 172. *See also* Libres

Labor: collective, 62–68 *passim*; and crop use, 72–73; nonagricultural, 73–75
Laguna region, settlement of, 8–9
Lagunero Indians, 9–10
Land: reform program, xii; irrigable, 6, 22, 38; tenure, 10–13 *passim*, 17–18, 22, 23–24, 37f, 165; size of holdings, 22–23; inheritance of, 39; borrowing and lending of, 57; rental of, 69, 179–80
Landless laborers, *see* Libres
League of Socialist Agronomists, 168
Leisure activities, 109–17 *passim*
Lerdo, Durango, 6
Lewis, Oscar, xii
Libres: limited rights of, 30, 50, 95–96; residence of, 32, 59, 89, 126; work of, 46, 62, 64, 66, 71, 75, 88; income of, 64, 66, 69, 86, 87–88, 143; status of, 89, 92–98 *passim*, 123
Literacy, 114
López Mateos, Adolfo, 151, 160

McBride, George, 12
Malinche, La, 106
Marriage, 16, 117–20
Matamoros, Coahuila, 25f, 60, 122, 132
Mechanization, xiii–xiv, 18, 58f, 69, 73, 109, 151
Membership, ejido, 39f, 89, 92, 178–79

Index

Mexican Revolution, xi, 12, 17
Mexico: Agrarian Department, 41, 50, 147f, 155; Water Commission, 6, 12. *See also* Federal government; *and individual presidents by name*
Michoacán, ejidos in, 153–54
Migration, 16, 35–36, 40, 92ff, 132f, 142, 175f
Mortality rate, 35, 175
Mosk, Sanford, 167
Mother's Day fiesta, 103

National Union of Rural Workers (CNC), xvi, 50–51, 147, 151, 170
Nazas River, xiv, 3ff, 9, 11, 22, 26, 54, 56

Official positions in ejido, 41–48 *passim*, 90f
Opinión, La, 8
Orona, Arturo, 51
O'Sullivan, J. S., 11

Parcela system, 39–40, 63ff, 69, 72, 86, 94, 156–57; and income variation, 57–58, 82f
PRI (Party of Revolutionary Institutions), xi, 51
Peones libres, see Libres
Population, 21, 32ff, 92, 132, 134, 175f. *See also* Migration
Political system: hacienda, 14; ejido, 40f, 47, 95, 169–70, 173
Profits, distribution of, 69, 83f
Property rights, 54–59 *passim. See also* Land
Purcell, Guillermo, 11

Religion, 16, 30, 104ff, 118–19, 125. *See also* Catholic Church
Remplases (substitutes), 40, 92–93, 126
Residence patterns, 15–16, 31f, 89, 124, 125–26
Ritual kinship, 16, 124–25
Ruiz Cortines, Adolfo, 52, 96, 160

Saltillo, Coahuila, 6f, 25, 48
San Isidro, fiesta of, 104, 110
San Pedro, Coahuila, 26
Santa María de las Parras, mission of, 10

School, ejido, 27–30, 99. *See also* Education
Secretary of ejido, 41, 43
Senior, Clarence, 150
Separatistas, 145, 146–48
Sharecropping, 153f
Siglo, El, 8
Sinaloa, 155
Sindicato, 36ff
Social activities, 30, 109–17 *passim*
Social classes, 90–91, 172
Social controls, 41, 71
Social Fund, 52, 61
Social mobility, 119, 144
Social relationships, 15, 119–20, 124, 174
Social services, 15, 157, 170
Society of Parents of Schoolchildren, 96, 102
Society of the Virgin of Guadalupe, 108
Socio delegado (official), responsibilities of, 41, 42–43, 47, 49
Specialized occupations, 73–75
Spending habits, 84–85, 121–22
State government, 19, 52
Status, 34, 45, 47, 89, 118; indices of, 90f, 113
Strikes, 19, 38

Tannenbaum, Frank, 13, 145
Taxes, 52
Technical assistance, 165–68 *passim*
Tepoztlán, ejido of, xii
Tlahualilo, ejidos of, 151–53
Tlahualilo Land Company, 11, 17, 26
Tlaxcalan Indians, 10
Torreón, Coahuila, 6, 9, 26, 116, 132
Transportation, 6–7, 8, 26
Treasurer of ejido, 41, 43f

Unionization, 18–19, 36ff, 96. *See also* Agrarian Committee; Central Union; CNC
United States, attitudes toward, 161
Urban influences, 8, 117
Urdiñola, Capt. Francisco, 10

Vigilance committee, 46–47
Virgin of Refugio, fiesta of, 104–5, 106, 110
Voz de Mexico, La, 52

Wages, 13f, 62–63, 79, 82, 84, 141
Water: supply, 3ff, 6, 26; rights, 11f, 22, 54, 55–56, 73, 172–73
Wheat, 4, 14, 63, 67ff, 77f
Whetten, Nathan, 13, 45, 85
Women, 13, 34, 40, 50, 108f, 116; status of, 120–21

Women's League, 121
Work chief of ejido, 44–46

Yaqui-Mayo area, Sonora, 149, 154f
Yucatán, 155

Zuloaga, Leonardo, 11